Evidence-Based Medical Ethics

Evidence-Based Medical Ethics

Cases for Practice-Based Learning

Authors
John E. Snyder, MS, MD

The University of North Carolina School of Medicine at Chapel Hill
SEAHEC (South East Area Health Education Center)
Department of Internal Medicine
New Hanover Regional Medical Center
Wilmington, NC, USA

Candace C. Gauthier, PhD

Department of Philosophy and Religion
University of North Carolina Wilmington
Wilmington, NC, USA

Foreword by
Rosemarie Tong, PhD

Department of Philosophy
The University of North Carolina at Charlotte
Charlotte, NC, USA

☀ Humana Press

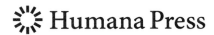

John E. Snyder
The University of North Carolina School
of Medicine at Chapel Hill
SEAHEC (South East Area Health
Education Center)
Department of Internal Medicine
New Hanover Regional Medical Center
Wilmington, NC, USA

Candace C. Gauthier
Department of Philosophy and Religion
University of North Carolina Wilmington
Wilmington, NC, USA

ISBN: 978-1-60327-245-2   e-ISBN: 978-1-60327-246-9
DOI: 10.1007/978-1-60327-246-9

Library of Congress Control Number: 2008923648

Cover illustration: Photo provided courtesy of Carlos Sillero.

Printed on acid-free paper

9 8 7 6 5 4 3 2 1

springer.com

Disclaimer

The purpose of this book is to help the reader learn about the principles of medical ethics by presenting fictional clinical cases that, in many ways, may mirror cases medical practitioners see in their daily practice. The specific cases described in this book are fictitious and any resemblance to real persons, living or dead, or real events is purely coincidental. The information herein is presented in good faith to be accurate and from reliable sources, based on available knowledge and information at the time of book publication. However, due to the possibility of human error, as well as the constant evolution of medicine and law, neither the authors nor the publisher contend that this book is complete in its scope of topical coverage, or accurate in every aspect relating to the ethical principles, medical practice, or laws described herein. The reader is encouraged to confirm the information in this book using available outside resources when possible, particularly when used to guide the care of actual patients. The authors and publisher disclaim responsibility for errors or omissions within this book, or for the results obtained from using the information herein. The laws discussed in this book are for illustrative purposes and may vary from state to state, or may have changed by the time of or after this book's publication. Practitioners should seek legal counsel when they are concerned about the legal consequences of their medical practice.

About the Cover

Justitia, the Roman goddess of justice, is often seen carrying a sword and scales and symbolizes fair and impartial officiation of the law. She acts without corruption, greed, prejudice, or favor. In recent centuries she is sometimes portrayed as "blinded" with a blindfold in order to suggest her objectivity more strongly. We have opted to use a traditional, unblinded face of Lady Justice for the cover of our book, as today's medical practitioners must face ethical dilemmas with open eyes and with consideration of health care justice for their patients. Solving ethical problems in patient care with a strong knowledge of facts—both medical and legal—and an awareness/acknowledgement of existing prejudices is essential in order to break down cultural barriers and offer the best possible care to all patients.

(Photo provided by Carlos Sillero with permission.)

Foreword

For more than 25 years I have focused my scholarly work and teaching efforts on issues that surface in the area of health care ethics/bioethics/medical ethics. Judging from the number of gray hairs on my head, I think it is safe to say that I have read nearly every book there is to read on topics such as competency, informed consent, end-of-life decision making, difficult patients, advance directives, medical error, distribution of scarce medical resources, and discharge of patients who have nowhere to go. But never have I read a book quite as compelling as John E. Snyder, MS, MD, and Candace C. Gauthier, PhD's book entitled *Evidence-Based Medical Ethics: Cases for Practice-Based Learning*. Although this talented team consisting of a physician (internal medicine) and health care ethicist have geared their book toward young medical trainees, their book — which combines reviews of evidence-based medicine, background in federal and state laws, and discussion of basic ethical principles — is equally well-suited to help members of institutional ethics committees learn how to use their critical reasoning skills and appropriate emotional intuitions to resolve common (yet pressing) ethical concerns in the context of practicing clinical medicine. Their book is also one that I am very eager to use as a teaching tool in the undergraduate and graduate health care ethics and law courses I teach.

What distinguishes Snyder and Gauthier's textbook from other books of its type is the way the authors force the reader to think through hard ethics cases and arrive at a resolution. So challenging are the cases that Snyder and Gauthier present, I kept hoping they had sent me a yet-to-arrive teacher's manual with the "answers" to the cases. But then I realized I did not want the "answers" to the cases. Instead I wanted to do the hard thinking that my new teachers, Snyder and Gauthier, had invited me to do. Congratulations to this duo for writing a book that I actually read cover to cover in one marathon session simply because each case in it so captivated my interest and imagination.

Rosemarie Tong, PhD

Preface

The idea for this textbook was born out of a need for a teaching resource that merges medical ethics theory with the practical needs of modern clinical medicine. Our goal in writing this book is to provide a method for the reader to learn how to systematically manage dilemmas seen in the everyday practice of medicine. The reader is guided through several "typical" patient scenarios and prompted by various questions that should be entertained by the treating health care provider. Then, relevant evidence-based medicine, legal precedent, and the ethical theory that applies to the situation are revealed. Often, finding the "best" ethical solution for each problem is automatic, as the solution often becomes self-evident during information gathering. This general method is reinforced throughout the text with multiple cases, using a practice-based approach by building on the reader's developing skills. Additionally, we have sought to emphasize a culturally competent manner to resolve these dilemmas, respectfully addressing issues of age, gender, and culture whenever possible. It is our hope that the reader will enjoy the mystery-solving style that each case offers and will subsequently adapt a patient-centered and evidence-based approach to the dilemmas they may face in their future practice of medicine or work in the medical ethics field.

John E. Snyder
Candace C. Gauthier

Contents

Foreword   ix

Preface   xi

Chapter 1   **An Introduction to Modern, Evidence-Based
Medical Ethics**   1

Chapter 2   **The Underlying Principles of Ethical Patient Care**   11

Chapter 3   **Advance Directives:** *The Living Will
and the Power of Attorney for Health Care*   17

Chapter 4   **Case-Based Ethical Dilemmas**   21

Case 1   **When Consent and Capacity Collide**   25

Case 2   **When a Patient's Health Care Agent Does Not Fulfill
Her Obligations**   31

Case 3   **When Coercion Dictates Care**   39

Case 4   **When a Spouse Is Estranged**   47

Case 5   **When a Patient Is Behaving Badly**   53

Case 6   **When a Patient Becomes Agitated**   59

Case 7   **When Patient Behavior Constitutes Abuse or Neglect**   65

Case 8   **When a Patient Has "Burned His Bridges"**   75

Case 9   **When a Patient Is Administratively Discharged**   81

Case 10 **When a Patient Makes Questionable Decisions** 87

Case 11 **When a Partner Is Excluded** 93

Case 12 **When a Diagnosis Is Reportable** 103

Case 13 **When Care Becomes Futile** 111

Case 14 **When Age Is a Factor in Health Care Decisions** 119

Case 15 **When a Patient Is Unidentifiable** 125

Case 16 **When Next-of-Kin Disagree** 131

Case 17 **When a Mistake Has Been Made** 139

Case 18 **When a Patient Codes** 145

Case 19 **When a Patient Places a Practitioner at Risk** 151

Case 20 **When Patient Non-Adherence Dictates Therapeutic Options** 161

Case 21 **When Family Members Limit Caregiver–Patient Communication** 167

Case 22 **When a Patient Requires Significant Medical Resources** 173

Case 23 **When a Patient's Belief System Affects His Care** 181

Case 24 **When a Minor Requires Confidential Medical Care** 187

Case 25 **When a Patient's Condition Is Terminal** 195

Comprehensive Exam 201

Comprehensive Exam — Answer Key 219

Glossary of Terms in Medical Ethics 233

Index 235

Chapter 1
An Introduction to Modern, Evidence-Based Medical Ethics

The Definition and Evolution of Medical Ethics

Ethics is a branch of philosophy centered on the study of the moral principles that define a standard of human conduct for an individual, group, or culture. The action of an individual or group, the motivation behind that action, and the end result of that action is often judged as being "good" or "moral" when scrutinized by the eyes of ethical philosophy. The choices of an individual, based on their personal moral beliefs, can have a consequential effect on other people or society at large. As a result, ethical standards are often applied to the general public in the form of policies and laws. Ethics can also be applied to various professions in order to define a level of responsibility or a standard code of performance for those in the field. The study of how the practice of medicine correlates with acceptable conduct is specifically termed *medical ethics*. National medical organizations have created ethical manuals to suggest guidelines for the ethical practice of medicine. Additionally, state medical boards that offer licensing for health care professionals also have regulations for obtaining and maintaining licensure based on standards of ethical practice. However, despite written directives for addressing common challenges, the practice of clinical medicine inherently will be fraught with unforeseen ethical dilemmas, since both the health care practitioner and the patient will bring their individual life experience and moral values to the clinical encounter. It is the job of the practitioner to find a balance between fundamental ethical principles, the law, and respect for the values of the patient.

On Medical Oaths, Codes of Ethics, and the Lessons of History

The Oath of Hippocrates was likely written in the Fifth Century B.C., and set forth a code of conduct for physicians to follow. Although still revered within its historical context, the Oath is vague regarding most of its implications for ethical clinical practice. Exceptions to this ambiguity include its specific support of respect for patient confidentiality and its strict prohibition of abortion, euthanasia, and sexual

From: *Evidence-Based Medical Ethics*
By: J.E. Snyder and C.C. Gauthier © Humana Press, Totowa, NJ

relationships between physician and patient. The Oath has been modified numerous times over the years and, since the early 1800s, it has been used in various forms at many medical school commencement ceremonies.

Some have contended that the Oath holds little relevance in modern medicine and that it is politically incorrect as it, for one example, refers to the medical care of slaves. It has been additionally argued that a one-time reading aloud of a doctrine, without contingent penalties for non-adherence to its principles, does not ensure the future ethical practice of any given practitioner. As a result, some institutions, such as the University of Massachusetts Medical School, have taken a different approach to the concept of the medical oath. The school holds an annual Oath Ceremony at the completion of the first two basic science years of training, just prior to entering the subsequent two years of clinical rotations. At the event students read an alternative oath of their own creation that reflects the scientific, ethical, and humanistic sides of medicine. Students are encouraged to develop an oath that states, in their own words, what "it means to be a physician" and "what sort of person you (*sic*) would like to be as a physician." The oath is reinforced at commencement exercises two years later, when students additionally recite the Oath of Maimonides – an edict which encourages physicians in its central theme to "never see in the patient anything but a fellow creature in pain."

The need for greater detail and clarity as to what ethical guidelines a physician must follow in practice – beyond a once-recited oath – has generated formalized codes of physician conduct. Some 24 centuries after the original Oath of Hippocrates, at the first meeting of the American Medical Association (AMA) in 1847, the organization's first Code of Ethics was approved. This Code sought not only to establish rules of professional behavior, but also to clarify the training requirements necessary to become a practicing physician. The Code of Ethics has since gone through several necessary revisions, largely reflecting changes in medical technology and in the social climate of this country.

Exactly one century later, in the Doctor's Trial of 1947 Germany, the indictment and prosecution of several Nazi physicians accused of human experimentation practices that included torture and murder, resulted in the tribunal's development of the Nuremberg Code. Largely relying on the principles of the Oath of Hippocrates and established standards of medical ethics at the time, the Nuremberg Code was a landmark document for promoting human rights protection in research. The concept of informed voluntary consent came forth from this document, as well as the right of the research subject to electively cease their involvement in a study at any time. All in all, the 10 decrees in the Nuremberg Code formed the basis for all modern day research regulations that protect the participants.

Since Nuremberg, and paralleling the equal rights movement and the increasing diversity of the population in our country, medicine has undergone a significant shift from a strictly paternalistic approach to one that embraces the rights of the individual as well as the concept of cultural competence. One certainly can argue, however, that this metamorphosis has been difficult at times, slow in evolution, and is by no means complete. Historical events, such as the Tuskegee Syphilis Study, have marred the reputation of health care in, possibly, an irremediable way. In this

U.S. Public Health Service study, spanning 40 years from 1932 to 1972, 399 African-American males from rural Alabama were given non-curative antibiotic courses for syphilis. The reported purpose of the study was infection control, and the prescribed antibiotic courses likely rendered most of the patients noninfectious. However, in the opinion of many experts, the study aimed not only to control disease transmission, but also to document the progression of the disease through its stages. Despite objections to the ethical premise of this investigation made as early as 1965, the study continued for another seven years before it was eventually deemed to have been unethical from the time of its inception by a nine-member advisory panel of the Department of Health, Education, and Welfare. The negative societal impact of this ruling has been immense. In the words of Dr. Vanessa Northington Gamble, the Tuskegee study "has come to symbolize racism in medicine, misconduct in human research, the arrogance of physicians, and government abuse of Black people."

Concerns for the protection of human research subjects, largely based on the issues raised by the Tuskegee Study, led to the formation of the National Commission for the Protection of Human Subjects in 1974, and its subsequent publication of *The Belmont Report: Ethical Principles and Guidelines for the Protection of Human Subjects of Research*. The Belmont Report led to the requirement that all research institutions receiving federal funding must form an Institutional Review Board (IRB) composed of individuals who review all research proposals from within the institution and ensure that the rights of the study subjects are protected. Members of IRBs now generally include both researchers and laypersons. IRBs review both proposals for research protocols, as well as the means by which informed consent is obtained for study subjects.

In 1979 Beauchamp and Childress penned the significant work *Principles of Biomedical Ethics*, now currently in its fifth edition. In this noteworthy and influential tome, the authors emphasize the importance of four basic ethical principles essential to medicine: autonomy, beneficence, non-maleficence, and distributive justice. These principles are discussed in greater detail in the following chapter and establish an important foundation for the ethical discussions throughout this text.

The fact that protection for research participants was born out of a history of human rights abuse must not be forgotten, and the work of physicians and scientists towards regaining the trust of minority groups is perhaps a Sisyphean task. However, encouraging minority group participation in research is essential so that data obtained is generalizable and all persons can reap the benefit of technological advances. Progress made in medical science, such as the introduction of hemodialysis and mechanical ventilation, have made it possible to prolong life in those who, in the past, would have succumbed to organ failure. However, such innovations have also created new ethical dilemmas. Questions as to who should receive such life-sustaining measures are often the most common received by palliative care and ethics consultants in the inpatient setting. To address such dilemmas, the American College of Physicians (ACP) has independently developed its own Ethics Manual that not only reviews basic tenets of medical ethics, but offers guidance on how to approach decision making in dilemmas faced in modern medical practice.

Medicine and the Law

Health care, as it is practiced today, has been influenced by the development of ethical codes and the lessons learned from egregious historical mistakes, as well as the evolution of health care law. Physicians are held ethically and legally accountable for maintaining set standards of care for their patients by law enforcement agencies, medical licensing authorities, insurance payers, and courts of law during malpractice litigation. One unfortunate consequence of the latter is the practice of so-called "defensive medicine," whereby a physician orders unnecessary studies on a patient in order to reduce the likelihood of litigation against them. Defensive medicine costs the health care system billions of dollars and likely does not help to decrease the number of lawsuits filed. The ethics of defensive medicine is dubious at best, considering the expense, time, and potential patient harm involved in its practice.

When a physician pursues training or a career in a new state, knowledge of the state medical board bylaws, as well as the governmental laws of the state itself, is crucial. A health care practitioner may or may not agree with existing policies, either for political, philosophic, religious, or other personal reasons. Yet awareness of their respective state laws is essential. For one example, elective abortion is currently legal in the United States, however individual states have varied restrictions on the procedure that are important for the practitioner to know. Some states have requisite waiting periods, others have mandatory pre-abortion counseling, and others have ordinances about parental notification and consent if the pregnant patient is a minor. Numerous states have regulations that prohibit insurance plans from covering abortion procedures. As of 2007, all but three states (Alabama, New Hampshire, and Vermont) allow individuals or health care facilities the choice to refuse a woman certain health services, information, or even referrals. A whole separate set of individual state regulations exist for providing patients accessibility to so-called "emergency contraception."

Health care law is in a state of constant evolution. Take, for example, the well-known case of Theresa Marie ('Terri') Schiavo [for an excellent review on the topic, please see Perry, et al.]. In 1990, at the age of 26, Mrs. Schiavo was determined to be in a persistent vegetative state (PVS) as a result of a cardiac arrest event and the associated hypoxemic brain injury. The cause of the cardiac arrest, although attributed by some to be due to electrolyte imbalances in the setting of an eating disorder such as bulimia, has never been officially confirmed. After exploring treatment options and pursuing aggressive rehabilitation therapies, Terri's husband came to the realization that his wife's condition was irreversible and that she would not want to continue living in a PVS. Terri's parents disagreed with this supposition, and felt that she showed outward signs of behavior that were not consistent with a PVS diagnosis. Years of trials and appeals in the Schiavo's home state of Florida ensued, ultimately resulting in the decision to withdraw supportive care. However, in the setting of widespread media attention and public interest, Florida governor Jeb Bush intervened by signing a policy known as Terri's Law – emergency legislation that resulted in the reinstitution of nutrition and other supportive care for Terri. When the Supreme Court of Florida deemed Terri's Law unconstitutional nearly a year later, her feeding tube was removed and she subsequently passed away at the age of 41. A postmortem examination was performed after Terri's death, and the brain pathology results were consistent with extensive past ischemic brain injury that included cortical blindness. Despite continued debate about whether the "right" decision was made for Terri, the Schiavo case clearly provides evidence that medicine, ethics, and law are closely intertwined and that significant fluidity of their interactions exists. Needless to say, strong knowledge of existing health care law, and the use of legal counsel when necessary, is becoming increasingly necessary in the practice of medicine.

Ethics in Medical Education Today

Ethics education is increasingly recognized as a crucial component in the training of medical professionals in the United States to teach health care practitioners to practice their trade and meet the obligations of an evolving ethical code in an increasingly complex practice environment. There has been a noticeable trend of increased emphasis on incorporating training in cultural competence in medical education as seen in the current accreditation guidelines of professional training programs. Take, for example, the education of an allopathic physician in medical school and residency. Since 1965, the Association of American Medical Colleges (AAMC), along with the Council on Medical Education of the AMA, have co-sponsored the Liaison Committee on Medical Education (LCME) which accredits the 125 United States and 17 Canadian allopathic medical schools. The following is excerpted from the LCME's guide to *Functions and Structure of A Medical School*:

> *"The faculty and students must demonstrate an understanding of the manner in which people of diverse cultures and belief systems perceive health and illness and respond to various symptoms, diseases, and treatments. Medical students must learn to recognize and appropriately address gender and cultural biases in themselves and others, and in the process of health care delivery. A medical school must teach medical ethics and human values, and require its students to exhibit scrupulous ethical principles in caring for patients, and in relating to patients' families and to others involved in patient care."*

Although education in ethics and professionalism is mandated by accrediting bodies, the manner in which this education is performed is left up to the school, and many different approaches exist. It is, perhaps, naïve to presume that, when a student of medicine proceeds through their professional education, the fundamental knowledge and skill in medicine they gain in school and postgraduate training is based solely on sound, unbiased science. The reality is that the training practitioner develops their clinical repertoire largely under the influence of role-modeling behaviors, both good and bad, that they observe by preceptors, colleagues, and other staff they interact with during their education. Medical school and residency programs increasingly recognize that education in medical professionalism and ethics, although classically taught through lectures, is perhaps better cultivated through mentoring relationships in the clinical setting by faculty role models. The University of Washington School of Medicine's Colleges Program is one example of a curriculum promoting the self-described "ecology of professionalism." The philosophy of this method is that strategic role modeling should occur throughout the duration of a UW student's medical school education.

Emphasis on formalized professionalism education is not limited to organizations overseeing the training of physicians. The National League of Nursing Accrediting Commission (NLNAC) and the Commission on Collegiate Nursing Education (CCNE) both emphasize ethics and cultural competence as major tenets of nursing education. Likewise, the American Council on Pharmaceutical Education (ACPE) sets professional standard education as a key paradigm of training for pharmacists. There is uniform recognition that respect for the patient and commitment to education of the fundamental ethical principles is paramount to the training and licensure of all health care professionals.

Cultural Competency, Personal Belief Systems, and the Practice of Ethical Care

The term *culture* best embodies the belief systems, attitudes, and behaviors that are characteristic of social, ethnic, age, geographic, religious, or other groups. In reference to clinical medicine, the term *cultural competence* reflects an ability to effectively integrate knowledge about cultural beliefs into positive therapeutic relationships, and to translate that into high quality comprehensive medical care. Instruction aimed at achieving cultural competence is increasingly being recognized as vital in medical education today as many students in medical professions may not have been previously exposed to persons of diverse backgrounds. Breaking down cultural barriers and increasing patient accessibility are essential to establish effective communication with, and provide overall care for, all patients. The health care practitioner must be aware of racial, ethnic, and other cultural differences when assisting patients and families through medical decision-making processes.

For example, a white physician caring for a terminally ill black patient with cancer may believe that the use of life-sustaining interventions such as mechanical ventilation would be futile, given the irreversibility of the patient's underlying medical condition. However, studies have shown that historical inequity in the care of black patients, such as what occurred in the Tuskegee experiments, has resulted in a perception among some black patients that limiting them from certain treatment options, such as a mechanical ventilator, is equivalent to an injustice by the medical system. An understanding on the physician's part that black patients may share such a cultural belief, and that this belief is a valid one, is crucial to maintaining open and effective communication with the patient and their family, and creating a successful therapeutic plan together.

It is equally important for the health care provider to self-reflect and recognize their own inherent cultural beliefs and how these may affect their ability to fulfill their clinical responsibilities. The topic of abortion, already discussed above, is one example of a politically charged topic that many health care practitioners are exposed to in training or thereafter. If a physician's religious views strictly forbid the practice of elective abortion, should that affect their ability to at least refer a woman to a physician that can perform the procedure? Should pharmacists with the same beliefs be allowed to not dispense the "morning after pill" to patients? Should a nurse provide a patient with unsolicited information about other options related to pregnancy, such as adoption? In addition to abortion, there are countless other examples of health care issues where a practitioner may have strong personal beliefs that may potentially influence their judgment. Attitudes towards assisted suicide, mandatory HIV testing in pregnancy, use of alternative therapies, and aggressive treatment in patients at the end of their life are just a few of such topics. Despite a diversity of cultural beliefs in health care workers themselves, their obligation must always be to their patients. Bridging differences across belief systems to provide patients with the best possible care remains the fundamental theme.

Health Care Providers and Conflicts of Interest

Influences beyond those related to culture can affect the choices made by a health care practitioner. One classic example is conflicts of interest created by relationships between providers and pharmaceutical companies. Clearly, these companies need to use marketing tools to maintain their profitability to further fund the research that may provide novel treatment options for the future. Billions of dollars are spent annually on advertising efforts by pharmaceutical companies. This includes both direct-to-consumer and direct-to-provider advertising. In a 2000 study by the National Institute for Health Care Management (NIHCM), about $2.5 billion was spent on direct-to-consumer advertising alone over the prior year. During this same period, the retail sales of the 50 medications most heavily advertised directly to consumers rose 32 percent (a $10 billion increase in retail spending), compared to a 13.6 percent sales increase for all other drugs combined. These top 50 drugs accounted for a total of $41.3 billion in sales in the year 2000, which comprised approximately 31.3 percent of the $131.9 billion spent overall on retail prescription drugs in that year.

Direct-to-physician marketing at medical offices, also termed "detailing," totaled $4 billion in this same 2000 study, which does not include the money spent on providing sample medications ($7.9 billion), hospital detailing ($765.3 million) and journal advertising ($484.4 million). Direct-to-physician promotions can include small gifts such as pens and notepads, but may also include dinners, travel funds, speakers and continuing education seminars, among other items. Worse yet, publications may be ghostwritten for physicians by compensated pharmaceutical company employees or consultants. One survey-based study by Blake and Early of patient attitudes toward gift giving by pharmaceutical companies to physicians suggests that, in general, patients disapprove of gifts that have nontrivial monetary value or do not benefit patients in some way.

It has been shown that direct commonplace marketing to medical students and residents by pharmaceutical sales representatives can influence their prescribing habits. Training these medical professionals to critically appraise information from pharmaceutical company representatives is essential to minimize prescribing biases and strengthen evidence-based practices. Interestingly, a study of internal medicine residents at the University of Toronto suggested that by implementing a policy that limits contact between them and pharmaceutical company representatives during training, residents were less likely to maintain a high level of contact with representatives in subsequent practice or to perceive information provided to them from representatives as beneficial.

In response to an increasing belief that the marketing by pharmaceutical companies influences the prescribing practices of physicians and physicians-in-training in a manner that is not based on empiric evidence alone, organizations such as No Free Lunch have formed to encourage health care providers to refuse promotional gifts. The American Medical Student Association (AMSA) has a similar campaign, titled Pharm-Free.

Ethics and Evidence-Based Medicine

Evidence-based medicine involves the thoughtful use of available and sound data from relevant clinical research to offer the best possible care for a patient during a specific encounter. The evidence-based practice of medicine offers more informed choices by both practitioner and patient, using the most studied diagnostic and therapeutic options available. However, such practice does not necessarily ensure that "better" choices, with regard to outcome or cost, are made. Additionally, the reality of decision making in medicine is perhaps best described in a paper by Kerridge, et al. and paraphrased here, as the culmination of knowledge from existing evidence and several other facets such as past experience, personal values, financial considerations, and concern for the relevant ethical principles. This interplay is just one example of the ways in which medicine is both a science and an art.

The above-referenced relationship between evidence and ethics in medical decision making is a bidirectional one. As much as depending on the fundamental principles of ethics can assist in making clinical determinations, evidence can help determine what decisions are most ethically sound. For example, the family of a patient with acute respiratory distress syndrome (ARDS) on a mechanical ventilator in an intensive care unit, is considering withdrawal of life-sustaining care. Based on the patient's previously stated wishes to withdraw life support in the setting of medical futility, knowing as much as possible about ARDS is essential. What is the usual mortality rate? In which patient population was this determined? What factors most affect mortality? What are the potential morbidities from the illness or treatments? What interventions have the most benefit? Would further aggressive measures truly be futile? What are the costs, potential benefits, and possible harms of these interventions? Avid use of the medical literature can assist the clinician in effectively guiding the patient's family through a very difficult decision. This recurrent theme, of using solid evidence-based medicine to help navigate through ethical decision making, with reliance on the basic ethical principles and established legal precedent, is strongly emphasized throughout this text.

References

Code of Medical Ethics (2002). American Medical Association.

Snyder L and Leffler C. Ethics Manual, Fifth Edition. Ethics and Human Rights Committee, American College of Physicians. Annals of Internal Medicine. 142(7):560–582, 5 April 2005.

A National Survey of U.S. Internists' Experiences With Ethical Dilemmas and Ethics Consultation. Duval, G SJD; Clarridge, B, PhD; Gensler, G MS; Danis, M MD. Journal of General Internal Medicine. 19(3):251–258, 2004.

Goldstein E et al. Professionalism in medical education: an institutional challenge. Academic Medicine. 81(10):871–876.

Ross J et al. Handbook for hospital ethics committees. American Hospital Publishing. 1986.

Graham D. Revisiting Hippocrates: does an oath really matter? JAMA. 284(22):2841–2842. 13 December 2000.

https://www.umassmed.edu/service.aspx?id=36192. Accessed October 15, 2007.

Oath of Maimonides. Acquired from: *http://www.fordham.edu/halsall/source/rambam-oath.html*. Accessed October 15, 2007.

The Nuremberg Code. Acquired from: *http://ohsr.od.nih.gov/guidelines/nuremberg.html*. Accessed October 15, 2007.

Shuster E. The Nuremberg Code: Hippocratic ethics and human rights. Lancet. Volume 351(9107), 28 March 1998, pp. 974–977.

Shuster E. Fifty years later: the significance of the Nuremberg Code. New England Journal of Medicine. 337(20):1436–40, 1997 Nov 13.

Northington Gamble V. Under the shadow of Tuskegee: African Americans and health care. American Journal of Public Health. 1997;87:1773–1778.

White R. Unraveling the Tuskegee Study of untreated syphilis. Archives of Internal Medicine. 2000; 160:585–597.

Seto B. History of medical ethics and perspectives on disparities in minority recruitment and involvement in health research. The American Journal of the Medical Sciences. 322(6): 246–250. November 2001.

Beauchamp TL and Childress JF. Principles of Biomedical Ethics, Fifth Edition, Oxford University Press, 2001.

Pro-Choice America. Who decides? The status of women's reproductive rights in the United States. Acquired from: *http://www.prochoiceamerica.org/choice-action-center/in_your_state/who-decides/maps-and-charts/*. Accessed October 15, 2007.

Blackston JW et al. Malpractice risk prevention for primary care physicians. The American Journal of the Medical Sciences. 324(4):212–219. October 2002.

Functions and Structure of a Medical School. Standards for Accreditation of Medical Education Programs Leading to the M.D. Degree. Acquired from: http://www.lcme.org. Accessed October 15, 2007.

Perry JE et al. The Terri Schiavo case: legal, ethical, and medical perspectives. Annals of Internal Medicine. 143(10):744–748. 15 November 2005.

Williams BA et al. Functional impairment, race, and family expectations of death. Journal of the American Geriatric Society. 54:1682–1687, 2006.

Blackhall LJ et al. Ethnicity and attitudes towards life sustaining technology. Social Science and Medicine. 1999;48:1779–1789.

Degenholtz HB et al. Race and the intensive care unit: Disparities and preferences for end-of-life care. Critical Care Medicine. 2003 Vol. 31, No. 5 (Suppl.):S373–S378.

"Culture." (n.d.). Dictionary.com Unabridged *(v 1.1)*. Retrieved July 29, 2007, from Dictionary. com website: *http://dictionary.reference.com/browse/culture*

Ladson GM et al. An assessment of cultural competence of first- and second-year medical students at a historically diverse medical school. American Journal of Obstetrics & Gynecology. 195(5):1457–62, 2006 Nov.

Achieving cultural competence guidebook from Administration on Aging, Department of Health and Human Services, United States. Acquired from: *http://www.aoa.gov/prof/adddiv/cultural/ CC-guidebook.pdf*. Accessed October 15, 2007.

Rudd KM. Cultural competency for new practitioners. American Journal of Health-Systems Pharmacy. 63:912–913. 15 May 2006.

Brennan TA et al. Health industry practices that create conflicts of interest. A policy proposal for academic medical centers. JAMA. 295(4):429–433. 25 January 2006.

Zipkin DA and Steinman MA. Interactions between pharmaceutical representatives and doctors in training. A thematic review. Journal of General Internal Medicine. 20:777–786. 2005.

McCormick BB et al. Effect of restricting contact between pharmaceutical company representatives and internal medicine residents on posttraining attitudes and behavior. JAMA. 286(16):1994–1999. 24/31 October 2001.

Blake RL and Early EK. Patients' attitudes about gifts to physicians from pharmaceutical companies. *The Journal of the American Board of Family Practice*. 8(6):457–464. November/ December 1995.

Stokamer CL. Pharmaceutical gift giving. Analysis of an ethical dilemma. Journal of Nursing Administration. 33(1):48–51. January 2003.

Prescription drugs and mass media advertising, 2000. A research report by The National Institute for Health Care Management (NIHCM) Research and Educational Foundation. Acquired from: *http://www.nihcm.org/~nihcmor/pdf/DTCbrief2001.pdf*. Accessed October 15, 2007.

No Free Lunch. *http://www.nofreelunch.org/*. Accessed October 15, 2007.

American Medical Student Association (AMSA) PharmFree Program. *http://www.amsa.org/prof/ pharmfree.cfm*. Accessed October 15, 2007.

Christiansen C and Lou JQ. Ethical considerations related to evidence-based practice. The American Journal of Occupational Therapy. 55(3):345–349. May/June 2001.

Sackett DL et al. Evidence based medicine: what it is and what it isn't. (it's about integrating individual clinical expertise and the best external evidence). British Medical Journal. 312(7023):71. 13 January 1996.

Chapter 2
The Underlying Principles of Ethical Patient Care

Medical ethics is based on a series of ethical principles that are particularly relevant to medical practice and patient care. The principles were first developed by Tom Beauchamp and James Childress in their 1979 book, *Principles of Biomedical Ethics*, now in its fifth edition. Since then they have become recognized and used by those who work in medical ethics, training medical students and residents, working as ethics consultants in health care institutions, and serving on hospital ethics committees. These principles are included in the "common morality," so they will be accepted by most members of the society. They are derived from classical ethical theories and are presupposed by traditional codes of medical ethics.

The principles of medical ethics make several contributions to patient care and decision making in the medical context. They offer a way to approach ethical dilemmas that arise in the course of practicing medicine, making difficult health care decisions, and interacting with patients and their families. The principles provide a way to organize our thinking about ethical issues in patient care and a shared language for health care providers to discuss these issues. Finally, the principles call attention to aspects of a medical situation that might be overlooked, and remind us that medicine is, ultimately, an ethical enterprise.

Beauchamp and Childress introduced four basic principles of medical ethics: Principles of 1) *Beneficence*, 2) *Non-Maleficence*, 3) *Respect for Autonomy*, and 4) *Justice*. In this book we have expanded the original list of four principles to include the *Principle of Respect for Dignity* and the *Principle of Veracity* to include elements of patient care and medical decision making that are not explicitly covered by the original four principles, such as emotions and relationships; privacy and bodily integrity; religious, social, and cultural differences; and the importance of good communication skills.

The Principle of Beneficence

Medical practitioners should act in the best interests of the patient.

More specifically, they should prevent harm, remove harm, and promote good for the patient. In the delivery of health care, the relevant harms to be prevented or

From: *Evidence-Based Medical Ethics*
By: J.E. Snyder and C.C. Gauthier © Humana Press, Totowa, NJ

removed may include pain and suffering, disease, disability, and death. Similarly, the relevant goods to be promoted may include well-being, health, proper functioning, and life. When applying this principle, it must be determined whether a proposed medical treatment will prevent or remove harm, or promote good for the patient. However, there may be disagreement about this. For example, while death is normally considered a harm to be prevented and life a good to be promoted, there may be medical situations in which this description is questioned. Family members may believe that the good for their loved one in a persistent vegetative state is continued life, even if this requires long-term artificial nutrition and hydration, and that death is the harm to be prevented. Health care providers, however, may believe that continued life, artificially maintained, is the harm to be prevented, while a natural death may be good for this patient.

The Principle of Non-Maleficence

Medical practitioners must not harm the patient.

This principle is based on the ancient maxim "First, do no harm" (*primum non nocere*). With the addition of this principle, the requirement to act in the best interests of the patient becomes more complicated. Again, disagreements may arise over the identification of harms for individual patients and specific medical interventions. In addition, medical interventions normally involve both harms and goods, often described as risks and benefits. This means that the *Principle of Beneficence* and the *Principle of Non-Maleficence* will often need to be applied together. Combining these two principles requires both an identification of the risks and benefits of a particular intervention and a comparison of the harms done, the harms prevented or removed, and the goods promoted for the patient. The result of these considerations will determine what is in the patient's best interests.

For example, when practitioners consider prescribing a particular medication, they must weigh the expected beneficial effects with the potential harmful side effects. As a result, providers can recommend the medication as being in the best interests of the patient, considering both the potential risks and the expected benefits.

In another example, a physician who writes a Do Not Resuscitate order for a frail elderly patient in the process of a natural death knows that this will result in the death of the patient. The physician in this case may believe that a natural death will not actually harm the patient but is, instead, a good to be promoted. Even if death is considered to be a harm, the practitioner has determined that CPR would cause more harm than the harm of death for this patient. Other health care providers, as well as family members, may disagree, considering the risk of broken ribs and other results of aggressive and invasive interventions to be less harmful than death.

The Principle of Respect for Autonomy

Capable patients must be allowed to accept or refuse recommended medical interventions.

"Autonomy" is defined as the capacity for self-determination or the capacity to make one's own decisions. In the health care context, this capacity involves the ability to make and communicate health care decisions. Respect for patient autonomy requires that those with this capacity be permitted to accept or refuse treatment alternatives recommended by their physicians. Of vital importance to the application of this principle is the requirement of *voluntary informed consent*. Capable patients must be provided with full, relevant, and truthful information about recommended treatments and any reasonable alternatives, including expected benefits, potential risks, and the results of refusing treatment altogether. They must understand this information and make a voluntary decision without coercion or undue influence.

Controversies arise here over the determination of who is capable of making these decisions. For many patients this will be obvious, based on their age or medical condition. One controversy involves mature minors, young people below the legal age of consent (18 years). Some have argued that the cognitive development of those who are 15 to 17 years of age qualifies them to make their own medical decisions. Another area of controversy involves those in the early and middle stages of Alzheimer's disease.

One solution is to evaluate the individual patient's capacity to make medical decisions, recognizing that these patients may be able to make some decisions and not be able to make others, depending upon the amount and difficulty of the information involved and the consequences of that decision. Those whose decision-making capacity is questionable should still be provided with information they can understand and be allowed to make age- and capacity-appropriate decisions.

The Principle of Respect for Dignity

Patients, their families and surrogate decision makers, as well as health care providers, all have the right to dignity.

The *Principle of Respect for Dignity* is meant to apply to everyone involved in the medical encounter. It is based on the fundamental idea that all persons should be treated with respect and dignity. Respect for persons and respect for their dignity applies whether or not health care decisions are being made, and even to those who are not capable of making their own decisions. Respect for people's dignity includes respect for their emotions, relationships, reasonable goals, privacy, and bodily integrity. Respecting these personal characteristics requires that they be acknowledged and taken into consideration in all medical encounters and in all aspects of patient care.

The *Principle of Respect for Dignity* applies to the relationships between practitioners and patients. It also applies to practitioners in their interactions with family members and surrogate decision makers. It requires respect for the social, cultural, and religious background of patients, their families, and surrogate decision makers. The principle reminds physicians that medical decisions are often made in the context of family and community background and history.

More specifically, this principle requires good communication skills, including active listening and the willingness to provide information, even when decisions are not being made. According to this principle, when family members and intimate friends accompany and support the patient, their emotions and their relationships with the patient should be acknowledged as valuable contributions to the patient's care. When surrogate decision makers are making decisions about medical treatment for a loved one, they need to receive the information necessary to make an informed decision, as detailed earlier. They should also be treated with the same respect for their emotions, their relationships, and their reasonable goals.

The *Principle of Respect for Dignity* also applies to patients who have not attained, are gradually losing, or have completely lost the capacity to make health care decisions. For example, children, the mentally disabled, and those with advancing Alzheimer's disease will experience emotions, participate in relationships, and have goals for themselves. All of these aspects of their lives should be acknowledged and taken into consideration in every encounter with them and in making treatment decisions. Their privacy and bodily integrity must also be protected to the extent possible, consistent with appropriate medical care and the need for decision making by others.

This principle also applies to those with severe dementia, comatose patients, and even those in a persistent vegetative state. They must always be treated with respect for their privacy and bodily integrity, as far as this is possible. Their inability to make decisions and even their inability to experience emotions and relationships must not allow medical practitioners to ignore their basic dignity as human beings.

The Principle of Respect for Dignity also requires confidentiality for patients' medical conditions and the treatments they are receiving.

This is an important element in the maintenance of the patient's privacy. Medical information about a patient must not be revealed to anyone who is not involved in the care of that patient. When patients are incapable of making their own medical decisions, information may be revealed to those who are legally authorized to make these decisions.

Medical practitioners must also preserve their own dignity in their encounters with patients, their families, and surrogate decision makers. They must approach these encounters with an expectation of respect and acknowledgement of their expertise. One of the ways in which this approach can be maintained is for practitioners to offer only those treatments they believe will be effective in meeting reasonable goals for the patient. For example, when family members request or demand that "everything be done," physicians need to be very clear, in their own minds, about what interventions will actually benefit the patient and what interventions will not. They should only offer those interventions they believe will be beneficial and explain why others, requested or demanded by the family, will not offer any benefit. Professional responsibility requires that physicians practice medicine according to their own judgment based on their training and experience, and not based solely on what family members want for their loved one.

The Principle of Veracity

The capable patient must be provided with the complete truth about his or her medical condition.

Capable patients must be provided with the complete truth about their medical conditions, both at the point of diagnosis and as their condition progresses. This is the only way that a patient can make a truly informed decision about accepting or rejecting recommended medical interventions. Patients must also be informed about their conditions in case experimental treatments were to become available. Similarly, surrogate decision makers must be provided with this information so that they can make an informed decision about the incapable patient's treatment.

Controversy arises when no medical treatment is available. Some argue that the diagnosis of a terminal illness or traumatic injury, with no available treatment, should be kept from the patient. Concerns about premature suicide or the loss of hope are often expressed. What is neglected in these arguments is that patients, as persons, have more to worry about than medical treatments. They have to consider their loved ones and what they can do for them now, before they die, and they have numerous other plans to make and see through before the end of their lives. Knowing the truth about what they can expect in terms of their illness or injury allows patients to make plans and live their lives with a purpose they may not have embraced before. Patients and surrogate decision makers also need to know what to expect as a terminal illness progresses so that informed decisions may be made about end-of-life treatment.

Both the *Principle of Respect for Autonomy* and the *Principle of Respect for Dignity* support the need for capable patients to be informed about their conditions, even if they are considered to be terminal, with no effective treatment. This is the only way that patients can make meaningful decisions about their medical treatment, how to respond to their emotions and relationships, meet their reasonable goals, and protect their privacy and bodily integrity as they make decisions about end-of-life care.

The Principle of Distributive Justice

Health care resources should be distributed in a fair way among the members of society.

The *Principle of Distributive Justice* is applicable when resources are expensive or scarce and decisions must be made about who will receive these resources.

The controversy with the *Principle of Distributive Justice* concerns what criteria will be used to distribute health care resources, so that the distribution is fair. Different theories of justice recommend different criteria for distribution, including ability to pay, merit, contribution to society, need, and first come, first served.

Our society seems to have adopted a combination of criteria for distributing health care resources. Patients must be treated in an emergency department for an acute illness or trauma. However, they may be released, once they are stabilized, if they do not have insurance or are unable to pay. Similarly, organ recipients must be on a list, using a first come, first served criterion, but must also be able to pay for the procedure and have adequate social support for the recovery process.

The *Principle of Distributive Justice* applies most readily on the governmental and institutional levels, in determining how much of our tax dollars will go to health care and in deciding how an institution's resources will be allocated. Yet, questions about the use of expensive and scarce medical resources may also occur to practitioners as they decide how to treat individual patients. Physicians may consult hospital administrators concerning these questions, but should not refuse medical treatment based on their own determination of the best use of resources.

Most of the principles presented above raise questions as they are applied to individual cases of patient care. Questions of what is harmful and what is beneficial, when harms outweigh benefits, benefits outweigh harms, or certain harms outweigh other harms, who is capable of making medical decisions, and when requested medical interventions are ineffective, must be answered by individual practitioners on a case-by-case basis. The principles may also conflict as they are applied to particular cases. For example, a capable patient may refuse a treatment that a physician judges to be more beneficial than harmful. On the other hand, legally authorized surrogate decision makers may demand medical interventions that physicians believe are ineffective and may even be harmful for the patient at the end of life.

Sometimes, when principles conflict, it is obvious which principle is most important and must take priority. At other times, however, practitioners are faced with an "ethical dilemma" in which the force of the conflicting principles seems to be equal. In these cases, it may be helpful to consult with the hospital ethics committee. This committee will consider the interests of everyone involved in the situation, as well as the relevant ethical principles, and may be able to recommend courses of action aimed at resolving the conflict. When conflicts seem intractable and involve legal concerns – for example, if practitioners are reluctant to provide clearly ineffective interventions demanded by surrogate decision makers – it may be necessary to include hospital administrators, risk management personnel, and hospital legal counsel in these discussions.

Annotated References/Further Information

Beauchamp TL and Childress JF. Principles of Biomedical Ethics, Fifth Edition, Oxford University Press, 2001.

Chapter 3
Advance Directives: *The Living Will and the Power of Attorney for Health Care*

When patients are conscious and capable, they must be permitted to make their own health care decisions, based on the *Principle of Respect for Autonomy*. However, the most difficult decisions are often made when a patient is no longer conscious or capable of participating in the decision-making process. This will be the case with most decisions about how patients' lives will end.

In 1976 California passed the Natural Death Act and became the first state to offer its citizens a way to make their wishes for end-of-life care known in advance. Adult patients suffering from terminal illness or mortal injury can execute a written document that authorizes withholding or withdrawing life-sustaining procedures. If physicians honor this "directive," they cannot be charged with criminal liability or unprofessional conduct. Other states followed with their own statutes authorizing what came to be known as the "Living Will."

The United States Supreme Court case *Cruzan v. Director, Missouri Department of Health* concerned the efforts of Nancy Cruzan's parents to have her feeding tubes removed after she had been maintained in a persistent vegetative state for seven years. The majority of the court decided that the parents would need "clear and convincing evidence" of Nancy's wishes before the feeding tubes could be removed. A Living Will, for example, would provide written evidence of a patient's wishes about end-of-life medical treatment. Justice O'Connor, writing in a concurring opinion, suggested that states consider authorizing "the patient's appointment of a proxy to make health care decisions on her behalf."

Following the Cruzan decision, many states added "persistent vegetative state" and "withholding and withdrawing artificial nutrition and hydration" to their Living Will statutes. Many states also allowed citizens to appoint someone to make medical decisions for them when they could no longer do so, often termed a "health care agent." Together the Living Will and the document used to appoint a health care agent have come to be known as "advance directives."

"Advance Directive" is the term used to refer to any document that makes the wishes of a capable patient clear for a time in the future when the patient is no longer able to make or communicate health care decisions.

When the patient is not capable of making health care decisions, as is often the case at the end of life, the *Principles of Respect for Autonomy* and *Respect for*

From: *Evidence-Based Medical Ethics*
By: J.E. Snyder and C.C. Gauthier © Humana Press, Totowa, NJ

Dignity may still be honored in the way these decisions are made and who is permitted to make them. Surrogate decision makers will be needed, for example, for those who are in a persistent vegetative state and those suffering from advanced dementia or the effects of a devastating stroke. The ideal surrogate decision maker would be a close family member, someone in a position to know or be able to make a reasonable determination of the patient's wishes.

It is possible to respect the wishes of a patient who can no longer make or communicate desires about medical treatment with the highest degree of certainty when the patient has an advanced directive that was completed when he or she was capable of doing so. The most common advance directives are the Living Will and the Power of Attorney for Health Care.

The Living Will expresses a patient's wishes that medical technology not be used to prolong the dying process.

The Living Will is a legally executed document authorizing physicians to withhold or withdraw life-sustaining medical treatment when the patient lacks the capacity to make health care decisions. Under most state laws the Living Will applies when the patient has a terminal condition or is permanently unconscious, as in a persistent vegetative state. However, some states limit the use of a Living Will to terminal conditions only, while others include conditions such as "serious disease or damage" to the brain (Maine), "seriously incapacitating illness or condition" (Missouri), and "end-stage conditions" including Alzheimer's and other types of dementia (District of Columbia, Maryland, and Virginia).

This first kind of advance directive is supported by the *Principle of Respect for Autonomy*. Based on a Living Will, practitioners can be reasonably confident that they are honoring their patients' wishes regarding life-sustaining interventions for certain medical conditions.

The Power of Attorney for Health Care is another form of advance directive.

The Power of Attorney for Health Care – in some states the "Durable Power of Attorney for Health Care" or "Health Care Power of Attorney" – is a legal document that a capable patient uses to appoint a health care agent (also termed "representative," "surrogate," or "proxy") to make medical decisions when he or she is no longer able to make or communicate such decisions. In most states the health care agent is authorized to request, receive, and review any medical information about the patient, including medical and hospital records; to employ and discharge health care providers; to consent to admission and discharge from health care institutions; and to give, withdraw, or withhold consent for diagnostic and treatment procedures.

Finally, the health care agent may legally authorize withholding or withdrawing life-sustaining medical interventions. The Power of Attorney for Health Care may also include a statement of specific desires for treatment at the end of life.

This second type of advance directive is supported by the *Principle of Respect for Dignity* and the *Principle of Respect for Autonomy*. With the Power of Attorney for Health Care, practitioners know who the patient appointed to make these decisions and, with the statement noted above, they will also know what the patient's wishes were regarding life-sustaining medical interventions.

Given the differences in advance directives from state to state, practitioners should familiarize themselves with the legally-authorized advance directives in the states where they practice medicine. Practitioners should also discuss advance directives with their patients before there is a need for end-of-life decision making.

End-of-life treatment decisions may also be made by legal guardians and family members in the absence of a Living Will or Power of Attorney for Health Care. When patients have not executed a Living Will or Power Attorney for Health Care, common law and state statutes permit these decisions to be made by family members. In most states the hierarchy for surrogate decision making, without a Power of Attorney for Health Care, is: legal guardian, spouse, and a majority of first degree relatives (parents and children). In 12 states, however, no priority is specified.

These legally-authorized surrogate decision makers must be fully informed of the patient's diagnosis and prognosis. They should also be encouraged to consider what the patient would want in terms of end-of-life treatment. Family members, in particular, may know what their loved one would want, even without an advance directive, based on conversations about these issues and verbally-stated desires. However, even without this kind of evidence, close family members may be able to reasonably determine what treatment would be desired based on the patient's values and beliefs (e.g., quality of life versus length of life and spending one's last days at home versus spending them in the hospital).

Surrogate decisions that rely on verbally expressed wishes or relevant values and beliefs are said to be based on "substituted judgment." This means that the surrogate is making a reasonable judgment intended to reflect, as closely as possible, the judgment the patient would make. The use of "substituted judgment" in surrogate decision making is justified by the *Principle of Respect for Autonomy*.

When surrogate decision makers have no basis on which to make a reasonable judgment about what the patient would want in terms of end-of-life medical treatment, they should be encouraged to consider the best interests of the patient. At this point, medical practitioners must fully inform family members about what the patient is experiencing and will experience as the relevant medical condition progresses. Family members may, again, be very helpful in making this kind of determination. For example, they are in the best position to know about the patient's tolerance for pain and total dependence on others. When end-of-life decisions are made in this way they are said to be made on the basis of the "best interests standard." Surrogate decision making that relies on the "best interests standard" is justified by the *Principles of Beneficence* and *Non-Maleficence*.

Advance directives are authorized by individual state legislatures, and physicians must educate themselves about the advance directives that are legally available in the states in which they are practicing medicine. It is also important that physicians discuss end-of-life treatment wishes with their patients when they are capable of considering them. Physicians should suggest that patients prepare whatever advance directives are legally authorized in their states, particularly if patients do not want their lives sustained in the event of certain medical conditions. In states where a Power of Attorney for Health Care is available, patients should be encouraged to appoint a health care agent to make medical decisions when they are no longer able to do so, and to discuss their wishes for end-of-life treatment with that person. This may reduce the conflicts that often arise between family members when no health care agent has been named.

Annotated References/Further Information

Cruzan, Natural Death Act, California Health and Safety Code, 1976. Part I, Division 7, Chapter 3.9, Sections 7185–7195. The California Natural Death Act was the first state law to authorize an advance directive.

Cruzan v. Director, Missouri Department of Health. United States [Supreme Court] Reports 497 (1990) 261–357. In this case, Justice O'Connor recommended that states allow patients to appoint a proxy to make medical decisions on their behalf.

Living Will Registry Website: *http://www.uslivingwillregistry.com.* Accessed October 15, 2007.

The Living Will Registry website provides links to the Living Will and Power of Attorney for Health Care for every state that has these advance directives.

Chapter 4
Case-Based Ethical Dilemmas

Learning Through Case-Based Teaching

In the university hospital, it is medical teaching round dogma that the best learning occurs when a student reads about a topic relevant to one of the patients they are caring for – reading it and seeing it shall permanently imprint the knowledge in the student's memory. David Irby, in his noteworthy article "Three exemplary models of case-based teaching" asserts that "building instruction around cases on the ward enables students and residents to see the direct relevance of the information to be learned, so that they are [*sic*] more highly motivated to learn it and are more likely to remember it."

In the spirit of this claim, the purpose of this book is to help the reader learn about medical ethics principles by presenting common clinical scenarios. Whether the reader is a graduate student of philosophy and ethics, a medical student or resident, a student in nursing or pharmacy school, or an active practitioner hoping to refresh or bolster their knowledge of medical ethics, it is the authors' hope that illustrating the principles of medical ethics with realistic cases of patient care will help solidify the reader's knowledge of ethics fundamentals, as well as provide a framework for approaching other ethical dilemmas they might encounter in the future. We also hope that seeing ethical decision making in action will instill an appreciation for the principles of ethics, and an enjoyment for finding ethical solutions to common challenges faced by health care practitioners.

How to Approach the Cases

Each of the following 25 cases is a fictionalized account of a common ethical dilemma seen by the health care practitioner in their daily work. Each case presentation is organized as follows:

From: *Evidence-Based Medical Ethics*
By: J.E. Snyder and C.C. Gauthier © Humana Press, Totowa, NJ

The Patient

The case is introduced with the patient's initial presentation to the practitioner, followed by a narrative account of their subsequent clinical course.

The Ethical Dilemma

The practitioner's ethical dilemma is presented. 'Questions for thought and discussion' are occasionally interjected, prompting the reader to pause and contemplate aspects of the ethical dilemma. These questions are particularly relevant for lecture-based discussions about the case.

The Medicine

A concise review of relevant medical literature is presented to offer a better understanding of the underlying data that supports different decision pathways in the case. Clearly the book does not seek to cover all of the evidence-based medicine behind the underlying diagnoses. However, information regarding standards of care, risk-benefit ratios, epidemiology, and cost-effective practices is presented to assist the reader in making determinations about the ethical dilemma at hand. The reference sources for the reviewed statistics, evidence-based medicine, and other information provided in each case are located in the Annotated References/Further Reading subsection. To learn more about each medical topic, seek the original articles as they are good learning resources. Newer information may be available since the publication date of this book, and the reader is encouraged to perform medical literature searches to ensure that they have the most up-to-date information at hand.

The Law

Some of the legal precedent relevant to the case, if established, is discussed. Note that laws vary state by state and are in constant evolution. The laws and rulings discussed in this book are for illustrative purposes only and the reader should seek legal counsel when they are concerned about the legal consequences of their medical practice.

The Ethics

The ethical principles of the case are discussed. Special emphasis is placed on the basic tenets of medical ethics: beneficence, non-maleficence, autonomy, justice, dignity, and veracity. Conflicts between legal precedent and ethically-appropriate choices are often highlighted.

The Formulation

In this subsection the reader is challenged to use the information from the previous subsections to create a therapeutic plan that is medically-sound, ethical, and follows existing legal precedent. With each case presented, a uniform framework of directions assists the reader in determining an appropriate patient care strategy. The goal of this subsection is to allow the reader to further develop and strengthen skills in solving ethical dilemmas. In 'The Formulation', a strong emphasis is placed on how to relate each case with the six "core competencies" of medical education (practice-based learning, medical knowledge, patient care, interpersonal skills and communication, systems-based practice, and professionalism).

Afterthoughts

Since patients with the same disease may present in infinitely different ways with regard to their symptom complex, so also can ethical dilemmas display variability. The 'Afterthoughts' subsection encourages the reader to consider additional dimensions of the ethical principles at hand by proposing alternative patient presentations. Additionally, the reader is often challenged to consider how patient characteristics such as culture, gender, race, ethnicity, and religious beliefs, among other factors, may affect practitioner decision making in the case.

Annotated References/Further Information

In every clinical case and associated dilemma presented, the reader is encouraged to seek further information, and there is a plethora of resources available in both written materials and via the internet. A select number of these resources, including those used to gather the data and information from the previous subsections, are provided at the end of each case.

Annotated References/Further Information

Irby, DM. Three exemplary models of case-based teaching. Academic Medicine. 69(12):947–953. December 1994.

Case 1
When Consent and Capacity Collide

The Patient

Mary B. is a 79-year-old female recently diagnosed with mild vascular dementia, who was living at home with her daughter, Audrey, and attending an adult day care program until she suffered a fall at home and was admitted to the hospital. Mary's past history also included hypertension, elevated cholesterol, and Type II diabetes mellitus – all of which were well-controlled by oral medications prescribed by her primary physician and that were administered by her daughter. Prior to her diagnosis of dementia two years prior to admission, Mary had lived by herself and functioned independently since being widowed at age 65. Most recently Mary was able to ambulate well at baseline, as well as bathe, dress, and feed herself. She had occasional episodes of mild confusion, particularly at night, but was for the most part oriented to person, place, and time. She did, however, have limitations with short-term memory, documented in prior mental status examinations. Mary had previously acknowledged in a Living Will that she did not wish to be resuscitated or intubated if she suffered a cardiopulmonary arrest, nor did she want a feeding tube placed if she became unable to support herself with oral nutrition. Mary had also designated Audrey as her health care agent in a prior Power of Attorney for Health Care document.

At the time of the current hospitalization, Mary had an unwitnessed fall at home while her daughter was in another room. When Audrey heard a commotion, she rushed to find her mother on the floor, with a small table turned over. Mary was, by Audrey's report, conscious at the time but slightly confused. Paramedics were called and Mary was brought to the emergency room. Upon arrival, Mary was no longer confused, but also had no recollection of the event. A medical evaluation done by the emergency room physicians determined that Mary had an intertrochanteric fracture of her right femur and a urinary tract infection. Additionally, a 12-lead electrocardiogram showed that Mary had developed third-degree heart block. She was admitted to the inpatient medical service for further evaluation and treatment, and the cardiology and orthopedic surgery teams were consulted.

From: *Evidence-Based Medical Ethics*
By: J.E. Snyder and C.C. Gauthier © Humana Press, Totowa, NJ

The Ethical Dilemma

The medical team caring for Mary, along with the respective consult teams from orthopedics and cardiology, recommended that Mary undergo surgical intervention for her hip fracture. Additionally, it was believed that Mary's fall was likely due to syncope from her heart block and that placement of a pacemaker was indicated to prevent further syncopal events. Mary stated to her medical team that she did not want hip surgery or a pacemaker. When the medical team attempted to illicit her reasons for refusing care, Mary became angry, refused to elaborate, and asked to be left alone. Mary's daughter, Audrey, tried unsuccessfully to convince her mother to undergo these procedures. Then, acting as Mary's health care agent, Audrey asked the medical team to proceed with both interventions despite her mother's protest, asserting that her mother was not competent to refuse treatment.

> **Question for thought and discussion:** How does one assess competency of this patient?

> **Question for thought and discussion:** Who should make the decision for performing the interventions in this case – the patient or her health care agent?

> **Question for thought and discussion:** When does forcing an incompetent patient to undergo treatment limit their right to dignity and go against the principle of non-maleficence?

The Medicine

Data presented after a congressional study in the matter (United States Congress OTA-BP-H-120) suggest that, among the over 300,000 persons admitted to United States hospitals for hip fractures annually, 94 percent are in people over the age of 50, and 55 percent are in people over the age of 80. Approximately 24 percent of patients over age 50 who have a hip fracture will die within one year after their fracture (all-cause mortality), with the highest rates of death occurring in males and in older patients. In females aged 75- to 84-years-old (like the patient in this case) all-cause mortality is 12 percent higher in those with a hip fracture in the previous year than those not sustaining a fracture. The majority of patients with hip fracture will undergo surgical interventions such as a pinning procedure or hip replacement surgery. About 10 percent of patients over 65-years-old will receive nonsurgical treatment of their hip fracture, and this approach is associated with worse overall patient outcomes, perhaps in part because these patients may be poor operative candidates. Even amongst patients who do undergo surgical intervention, most experience at least some degree of functional impairment, compared to their pre-fracture state.

Placing a dual-chamber pacemaker in a patient such as Mary (one with acquired, symptomatic third-degree atrioventricular block) is a Class I recommendation by ACC/AHA/NASPE 2002 guidelines (there is evidence and/or general agreement that a given procedure or treatment is useful and effective). One study (Rinfret, et al.) estimated that a 74-year-old patient undergoing implantation of a DDDR pacemaker would incur $59,104 in lifetime costs, and would have a life expectancy of 6.49 years.

The Law

The Power of Attorney for Health Care goes into effect when the patient is unable to make health care decisions. This is variously described as being "incapacitated," "incapable" of making these decisions or "lacking the capacity" to make them, in different state laws. Some states also include the inability to communicate health care decisions in this description.

The physicians in this case need to determine whether or not Mary is capable of deciding whether to accept or refuse the recommended procedures. It would make sense to ask Mary's primary care physician to talk with her and provide an opinion on her capacity to make this decision. Her own physician is likely to have insights based on their ongoing relationship that may be helpful in assessing her present capacity.

The medical team should keep in mind that a patient may be capable of making some decisions, but not capable of making others. This may depend, in part, on the complexity of the information needed to make a particular decision. In determining Mary's capacity to make the treatment decision in question, it will be important to ensure that Mary is fully informed about the recommended procedures and that she has understood this information. After Mary has been fully informed, questions should be asked about the information provided. If she is unable to demonstrate sufficient understanding with her answers, then she is probably not capable of making this decision.

If Mary understands the risks and benefits of these interventions and what her quality of life will be without them and still refuses, this should not be taken as definitive of her incapacity. At this point, efforts should be taken to elicit her reasons for refusing recommended treatment, including calling upon other hospital resources. For example, Mary may feel more comfortable having this conversation with a social worker or a chaplain.

Mary's reasons for refusing hip surgery and the pacemaker will be very important in determining her capacity to make this decision. Her reasons may make sense and may be convincing, leading the medical team to believe she is capable of decision making in this case. If not, they may indicate her lack of capacity to decide about these procedures. Finally, if the attending physician cannot determine capacity, a psychiatric consult could be called to provide another level of expertise.

If Mary is ultimately found to be incapable of making this particular health care decision, her daughter, as health care agent, would have the legal authority to consent to the recommended procedures. Mary's Living Will would *not* apply in this case because it refers to life-sustaining procedures when the patient is terminally ill or permanently unconscious.

The Ethics

According to the *Principle of Beneficence,* physicians should act in the patient's best interests. In this case the medical team believes the recommended procedures are in Mary's best interests. Certainly, repairing the hip fracture is necessary for the benefit of Mary's continued mobility, and the pacemaker is necessary to prevent the harm of more blackouts and traumatic falls. The patient's quality of life is also relevant here. Not repairing the hip fracture would likely condemn Mary to life in a wheelchair.

When the *Principle of Non-Maleficence* is considered along with the *Principle of Beneficence*, the risks and benefits of a particular medical intervention should be compared. In this case, the medical team believes that the benefits of hip surgery and the pacemaker outweigh the risks of these procedures.

According to the *Principle of Veracity*, Mary should be given complete and truthful information about her condition, regarding her hip fracture and heart block, in language she can understand. She needs to be told what these conditions mean for her future health, well-being, and quality of life.

Based on the *Principle of Respect for Autonomy*, capable patients must give their voluntary informed consent for medical treatment before it is undertaken, and they are also allowed to refuse recommended treatment. Because of this, an initial refusal of recommended interventions should not be taken as indicative of incapacity, as Mary's daughter has suggested. If a patient's capacity is questioned, as in this case, the consent process should still be attempted as part of the effort to determine that capacity. The consent process involves providing information, checking for comprehension, and seeking agreement from the patient, without coercion or undue pressure.

In this case, the medical team should present the risks and benefits of the recommended procedures and explain why they believe the benefits outweigh the risks. They should also describe any viable alternatives and the expected outcome if Mary refuses these procedures. She needs to know what her quality of life will be without these interventions: not being mobile, with more blackouts due to her heart block, and possibly future falls and physical injuries. These descriptions and explanations should be in language appropriate to Mary's level of understanding. By asking relevant questions the medical team should determine whether or not Mary has understood the information presented to her and the reasoning behind it.

The "voluntary" element of informed consent may be questioned in cases like Mary's since the risks of refusing the recommended procedures may create emotions of fear and anxiety in the patient that could be perceived as "undue pressure." However, as long as the information provided is accurate and presented in a calm, professional manner, without efforts to heighten the emotional reaction, this requirement can be satisfied.

After the informed consent process, if Mary refuses the recommended procedures, her reasons for refusing should be explored to determine if she truly comprehends the information provided. Her reasons for refusing may also be indicative of her capacity to make this decision, if they are understandable reasons, or her incapacity, if they are not.

According to the *Principle of Respect for Autonomy*, if Mary is found to be capable of decision making in this situation, she should be permitted to refuse the recommended procedures. If this occurs, alternatives that could reduce the harmful consequences of forgoing these interventions and improve her quality of life should be explored.

Whether Mary is capable or incapable of making her own medical decisions, she must be treated with dignity. The *Principle of Respect for Dignity* requires the medical team to consider Mary's history, belief system (for example, religious background), and her own goals for medical treatment. These may be learned from eliciting her reasons for refusing the recommended procedures. If her reasons are based on her history or religious beliefs, that would help the medical team in understanding her decision. If Mary's goals for treatment are reasonable, such as avoiding pain and dependence on others, they should be considered in determining her capacity and in making treatment decisions for her, if needed.

Mary's emotional response to this medical crisis must also be acknowledged and considered. Her initial refusal and anger may need to be explored, perhaps by a social worker or chaplain. Her anger may be a response to fear and anxiety in the face of a future of debilitation and dependence. There may also be relationship issues in this case. Mary may feel that she is a burden on her daughter and this feeling may be motivating her refusal of medical treatment.

Privacy will be hard to protect in this situation, as the circle of those who are involved in this case expands. However, one way to safeguard Mary's privacy would be to hold conversations with the medical team members, social worker, chaplain, and psychiatrist in a private setting, without the daughter present, at least while her capacity is being determined. This may also allow Mary to express relationship concerns that may be influencing her refusal of the recommended procedures. If Mary is found to be incapable of making her own decisions, her privacy and bodily integrity must be respected as far as possible, consistent with fully informed decision making by her daughter and recommended medical interventions to which the daughter consents.

The Formulation

Now that the evidence-based medicine, legal precedent, and relevant ethical principles for this case have been reviewed, formulate a strategy to address the ethical conflicts in this case. If necessary, perform additional research into local and state laws and hospital regulations. Consider delving further into the background medical literature to assist with making sound therapeutic decisions. Devise a treatment approach that addresses the needs of the patient and her family, that is both ethically and medically sound, and that is culturally competent. Ensure that the strategy employs fair and appropriate utilization of medical resources, and that the approach is practical and feasible within the limits of the medical system at large. Work out a clear and professional way to communicate the proposal to the patient and her family. Attempt to foresee challenges that may arise in conveying or implementing the plan. Determine what follow-up will be necessary to ensure that the chosen strategy remains successful for the patient in the long-term. Reflect on how the knowledge and skills learned from this case can be used to improve the care of patients that may be encountered in future practice.

Afterthoughts

In this case, ethical conflict arose because a patient refused treatment that others thought was in her best interest, and there was a claim that the patient was not competent to decide to refuse this care. Competency becomes a complex issue when a patient has a high level of functioning overall, but is limited in certain skills required for advanced decision making. Would this case have been different if the patient was diagnosed with a mental illness other than dementia, such as major depression or schizophrenia? What if the patient was an adolescent? A young child?

Annotated References/Further Information

Code of Medical Ethics of the American Medical Association: Current Opinions with Annotations, 2006–2007 Edition. Council on Ethical and Judicial Affairs. Annotations prepared by the Southern Illinois University Schools of Medicine and Law. The American Medical Association has published several opinions on ethical dilemmas relevant to this case presentation in their Code of Medical Ethics. Of note are the subsections on Informed Consent (8.08), Surrogate Decision Making (8.081), and Ethical Responsibility to Study and Prevent Error and Harm (8.121).

Snyder L, JD, and Leffler C, JD. Ethics Manual, Fifth Edition. Ethics and Human Rights Committee, American College of Physicians. Annals of Internal Medicine. 142(7):560–582, 5 April 2005. The Ethics Manual of the American College of Physicians (Fifth Edition) addresses several of the issues at play in this case in the subsections of Informed Consent, Making Decisions Near the End of Life, and Advance Care Planning.

Living Will Registry Website: http://www.uslivingwillregistry.com. Accessed October 15, 2007.

Braithwaite RS et al. Estimating hip fracture morbidity, mortality, and costs. Journal of the American Geriatrics Society. 51(3):364–70, 2003 March.

United States Congress, Office of Technology Assessment, Hip Fracture Outcomes in People Age 50 and Over-Background Paper, OTA-BP-H- 120 (Washington, DC: U.S. Government Printing Office, July 1994).

Rinfret S et al. Cost-effectiveness of dual-chamber pacing compared with ventricular pacing for sinus node dysfunction. Circulation. 2005;111:165–172.

Gregoratos G et al. ACC/AHA/NASPE 2002 guideline update for implantation of cardiac pacemakers and antiarrhythmia devices: summary article: a report of the American College of Cardiology/American Heart Association Task Force on Practice Guidelines (ACC/AHA/NASPE Committee to Update the 1998 Pacemaker Guidelines). Circulation. 2002;106:2145–2161.

Case 2
When a Patient's Health Care Agent Does Not Fulfill Her Obligations

The Patient

Henry C. is an 84-year-old male with severe Alzheimer's disease and multiple other chronic medical problems who was living in the dementia unit of a nursing facility when he was admitted to the hospital with fever, cough, and labored breathing. Henry was a widower without living children, who was functioning independently at home until his diagnosis of Alzheimer's dementia eight years prior to admission. His only known next of kin was a nephew who lived distantly, and with whom Henry had infrequent contact.

When Henry had had only early signs of mild cognitive impairment, he had completed legal paperwork to assign a health care agent – and chose a neighbor and friend, Jean H., to fulfill this role. Jean was also designated as Henry's agent in his Power of Attorney, charging her with managing his finances and other non-medical affairs were he to become incapable of doing so himself. As Henry's cognition declined and he became incapable of living safely on his own, he was admitted to the nursing facility for his long-term care needs. It was at this time that Jean began to manage Henry's finances and other non-medical needs. In the subsequent years Henry only had intermittent, mild acute illnesses at the nursing facility. All of these were managed without hospitalization until the current illness. However, his neurological status continued to decline to the point where he became minimally conversive, had significant memory impairment, and became chronically incontinent of urine and stool.

At the time of the current admission, a nurse had come into Henry's room at the nursing facility and noted he had labored breathing, a deep cough, worsening of baseline confusion, and a temperature of 102.3°F. He was brought by ambulance to the hospital where he was found to be in mild respiratory distress. A reliable medical history could not be obtained from Henry himself as he was confused and answered all questions with the word "yes." By exam, laboratory data, and radiography he was diagnosed with severe pneumonia and mild renal failure, the latter likely from volume depletion. He was admitted and started on routine treatment that included broad antibiotic coverage and IV fluids.

From: *Evidence-Based Medical Ethics*
By: J.E. Snyder and C.C. Gauthier © Humana Press, Totowa, NJ

The Ethical Dilemma

The medical team attempted to contact Jean by phone several times on the night of admission to give her an update on Henry's medical condition and to determine if Henry had completed a Living Will that specifically addressed his wishes were he to clinically decline or have a cardiopulmonary arrest. Two days after admission Henry's condition showed little improvement, and Jean had not been successfully contacted. On the third day of admission Henry's condition worsened. His breathing became more difficult and he required greater amounts of supplemental oxygen. Additionally he was noted to have mildly elevated serum troponin levels, suggestive of myocardial strain. He was transferred to the medical intensive care unit (MICU) for closer observation. The MICU team successfully reached Jean by phone later that evening, who reported that Henry had not specifically communicated his desire for or against heroic efforts in the setting of terminal illness or cardiopulmonary arrest with her. She told the physicians that she was "too old" to come to the hospital to visit Henry and to "just keep doing what you're doing." She denied knowledge of Henry's health problems other than "dementia." She also stated that she did not know if Henry had any living relatives that may want to be notified of his poor clinical status.

> **Question for thought and discussion:** As Henry's designated health care agent, is Jean adequately fulfilling her role?

The MICU team called Henry's nursing facility to determine if he had any other emergency contacts and received the contact information for Henry's nephew, Jim. Additionally they learned that Jean had not submitted payment for some of Henry's expenses at the nursing facility for several months and that this matter was now in the hands of the facility's legal department. Henry's nephew was called and was surprised to hear of Henry's poor medical condition. Jim stated that he had met Jean on several occasions and had concerns that she did not have Henry's best interests in mind. He felt that Jean was "only interested in my uncle's money." Although Jim admitted not having seen Henry in person in almost three years, he felt strongly that if Henry were able to express his wishes himself, that he would not want his life prolonged by mechanical ventilation or other means of Advanced Cardiovascular Life Support (ACLS) in the event of a cardiopulmonary arrest. He stated that Jean "probably wants to keep him alive so she can still collect his checks."

> **Questions for thought and discussion:** What is the obligation of the medical team to investigate Jean's capacity to act in Henry's best interest as his health care agent? Is there already enough information to determine that she is not doing so?

> **Questions for thought and discussion:** If Jean is determined to not be acting in Henry's best interest, who becomes Henry's surrogate decision maker? Is Jim an appropriate choice? Who determines this?

Question for thought and discussion: What happens if Henry has a cardiopulmonary arrest before it is determined who is best to act as his surrogate decision maker – is the medical team obligated to perform ACLS protocols, including placing Henry on a mechanical ventilator?

The Medicine

Using a Pneumonia Severity Index Calculator, an 84 year old male nursing home resident with cerebrovascular disease, renal failure, altered mental status, and hypoxemia is considered a "high risk" patient and has an estimated 30-day mortality of at least 26.7% on admission. Kaplan et al., in a study of 150,000 elderly Medicare patients with an inpatient diagnosis of pneumonia, determined that one-third who survive initial hospitalization will die within one year of discharge (all-cause mortality), most of these within the first three months.

Data regarding the effectiveness of in-hospital resuscitation of patients with cardiac arrest show variable results. In one ten year study of 732 consecutive VA patients with cardiac arrest (Bloom HL et al.), only 6.6% survived until hospital discharge. Those surviving until discharge had a 41% survival rate at three years; however this number was 77% in patients with implanted cardioverter-defibrillators. The American Heart Association states on their website "Early CPR and defibrillation [*sic*] within the first 3–5 minutes after collapse, plus early advanced care can result in high (greater than 50 percent) long-term survival rates for witnessed ventricular fibrillation."

In a study of older patient's views on resuscitation at the end of life (Somogyi-Zalud et al.), the majority (66%) of study patients preferred comfort care in the last six months of their life, and desired a Do Not Resuscitate (DNR) order in the last month of their life. Interestingly, another study of 70 patients (Seckler et al.) suggested that substituted judgment by a surrogate decision maker often did not accurately reflect a patient's wishes regarding desire for resuscitation. Although the majority of patients in this study predicted that their family members (87%) or their physician (90%) would accurately guess their preferences for resuscitative efforts, only 16% of family members and 7% of physicians were actually able to do so.

The Law

Henry's health care agent, Jean, is not willing (or able) to come to the hospital to receive complete information about his medical condition and recommended treatment options so that she can participate in decision making for him. For this reason she cannot act as Henry's health care agent.

Most state advance directives include a statement to the effect that if the person who is appointed as the health care agent is unwilling or unable to serve or is unavailable, an alternative can be named in the document. Thus, in these states, the law recognizes that a person may be appointed as health care agent, but may not be able to act in that capacity when the time comes.

In this case Henry did not name an alternative. However, according to common law, one of the patient's family members could act as surrogate decision maker. Since there are no other living relatives, Henry's nephew, Jim, would be the natural choice to become his surrogate decision maker. It would be best, however, for the medical team to consult with the hospital's legal department before dismissing Jean as health care agent.

If Jean had been willing to come to the hospital and participate in making treatment decisions for Henry, she may have made decisions the medical team believed were not in Henry's best interests, perhaps for financial reasons of her own. In this case, the medical team would need to consult the legal department of the hospital for advice on how to override Henry's Power of Attorney for Health Care. Different states and health care institutions are likely to have different policies about this.

If Jim does replace Jean as surrogate decision maker for Henry, Jim needs to come to the hospital to meet with the medical team and become fully informed about his uncle's diagnosis and prognosis, as well as the recommendations the medical team has for his treatment.

The question of whether or not Henry should be resuscitated and placed on a ventilator, if he has a cardiopulmonary arrest before the determination can be made about a surrogate decision maker, is a difficult one. Henry does have a legally-appointed health care agent who, ideally, should make this decision. However, Jean claimed not to know what Henry's wishes were for a case like this and she did not specifically request that he be resuscitated. The only other person available to make this decision (Jim) said his uncle would not want his life prolonged in this way. It could be argued that, based on the responses given by both of the possible surrogates, Henry should not be resuscitated and placed on the ventilator.

The Ethics

First, the physicians on the medical team need to decide what treatment they would recommend for Henry, given his deteriorating condition and advanced Alzheimer's disease. They may believe that, in the case of a cardiopulmonary arrest, it would be in Henry's best interests not to be resuscitated and placed on a ventilator. If so, they should be prepared to explain their recommendations to Henry's surrogate decision maker, in terms of risks and benefits for Henry, using the *Principles of Beneficence* and *Non-Maleficence*.

Jean has a legal claim to be Henry's surrogate decision maker, since she was appointed as his health care agent. Allowing her to make decisions about Henry's treatment would appear, on first consideration, to honor the *Principle of Respect for Autonomy*. However, Jean is unable or unwilling to serve in this role. For example, she doesn't seem able to make decisions for Henry based on "substituted judgment," since she doesn't appear to know much about what Henry would want. She also doesn't seem capable of making these decisions in his best interests, since she is either unable or unwilling to come to the hospital to be fully informed and participate in the decision-making process.

Henry's nephew, Jim, seems better suited to serve as the surrogate decision maker, in this case. He believes that his uncle would not want life-sustaining procedures, so he is thinking about what his uncle would want in this situation. He also seems to care about his uncle more than the health care agent does, so he may be more willing to consider what would be in his uncle's best interests. If the medical team decides that Jim is a more appropriate surrogate, he should be asked to come to the hospital to receive detailed information about his uncle's diagnosis and prognosis, as well as the recommendations of the medical team. That way he can actively participate in the decision-making process for his uncle.

If the medical team determines that CPR and a ventilator would not be effective in meeting any reasonable goals for Henry (for example, because they will only prolong life without restoring his cognitive capacity) they should not offer to initiate these procedures. They should explain to Henry's surrogate decision maker why they believe these interventions would be ineffective for Henry's condition. They should explain what measures they are willing to take to make Henry comfortable and to relieve any pain he may be experiencing. Based on the *Principle of Respect for Dignity*, physicians should not be pressured into providing medical interventions they believe provide no overall benefit to their patients.

The *Principle of Respect for Dignity* also applies to Henry. Even if he cannot make his own health care decisions or communicate effectively, Henry must still be treated with respect for his privacy and bodily integrity, as far as this is possible. For example, in their conversations with his surrogate decision maker, the medical team could point out that CPR and mechanical ventilation are both invasive procedures that will not stop Henry's physical and mental decline.

Because a health care agent was appointed and there is also a relative willing to make medical decisions for him, it could be argued that the medical team should resuscitate Henry and place him on a ventilator, if necessary, until a surrogate decision maker can be identified. This would allow the surrogate time to make a well-informed and well-considered decision. It would then to be possible to remove the ventilator, if that was the surrogate's decision.

However, the medical team may believe CPR and other life-support measures would be futile for Henry, and may cause him harm without providing any overriding benefit. If so, the medical team would not offer these procedures to Henry's surrogate decision maker, so they will not be options for the surrogate to consider. If this is the case, Henry would not be resuscitated or placed on the ventilator, even with a surrogate available and involved in making decisions about Henry's end-of-life care. For this reason, it may not be necessary to employ these procedures before a surrogate decision maker is identified.

The Formulation

Now that the evidence-based medicine, legal precedent, and relevant ethical principles for this case have been reviewed, formulate a strategy to address the ethical conflicts in this case. If necessary, perform additional research into local and state laws and hospital regulations. Consider delving further into the background medical literature to assist with making sound therapeutic decisions. Devise a treatment approach that addresses the needs of the patient and his family, that is both ethically and medically sound, and that is culturally competent. Ensure that the strategy employs fair and appropriate utilization of medical resources, and that the approach is practical and feasible within the limits of the medical system at large. Work out a clear and professional way to communicate the proposal to the patient and his family. Attempt to foresee challenges that may arise in conveying or implementing the plan. Determine what follow-up will be necessary to ensure that the chosen strategy remains successful for the patient in the long-term. Reflect on how the knowledge and skills learned from this case can be used to improve the care of patients that may be encountered in future practice.

Afterthoughts

In this case, Henry became unable to express his wishes with regard to supportive medical care at the time of critical illness due to cognitive impairment. Although he had appointed a health care agent to assist with these decisions early in the course of his dementia, he did not explicitly describe his wishes in a Living Will document. Additionally, the primary medical team had concerns that his chosen health care agent was not working in his best interests. To complicate matters, Henry's only surviving family member had minimal contact with him in the preceding years. Would this case have been different if Henry had no surviving family members? In how much depth do you feel a medical team needs to investigate the ulterior motives of any patient's health care agent or surrogate decision maker? Do questions need to be asked about these motivations even when there are no obvious conflicts of interest?

Annotated References/Further Information

Code of Medical Ethics of the American Medical Association: Current Opinions with Annotations, 2006–2007 Edition. Council on Ethical and Judicial Affairs. Annotations prepared by the Southern Illinois University Schools of Medicine and Law. The American Medical Association has published several opinions on ethical dilemmas relevant to this case presentation in their Code of Medical Ethics. Of note are the subsections on Informed Consent (8.08), Surrogate Decision Making (8.081), and Ethical Responsibility to Study and Prevent Error and Harm (8.121).

Snyder L, JD, and Leffler C, JD. Ethics Manual, Fifth Edition. Ethics and Human Rights Committee, American College of Physicians. Annals of Internal Medicine. 142(7):560–582, 5 April 2005.

The Ethics Manual of the American College of Physicians (Fifth Edition) addresses several of the issues at play in this case in the subsections of Informed Consent, Making Decisions Near the End of Life, and Advance Care Planning.

Fine MJ et al. A prediction rule to identify low-risk patients with community acquired pneumonia. New England Journal of Medicine. 336:243–250. January 23, 1997.

Pneumonia severity index calculator. December 2003. Agency for Health Care Research and Quality. Rockville, MD. *http://pda.ahrq.gov/psicalc.asp*. Accessed October 15, 2007.

Kaplan V et al. Pneumonia: Still the old man's friend? Archives of Internal Medicine. 2003;163:317–323.

Somogyi-Zalud E et al. Elderly persons' last six months of life: findings from the Hospitalized Elderly Longitudinal Project. Journal of the American Geriatrics Society. 48(5 Suppl):S131–9, 2000 May.

Seckler AB et al. Substituted judgment: how accurate are proxy predictions? Annals of Internal Medicine. 1991;115:92–98.

Cummins RO et al. Recommended guidelines for reviewing, reporting, and conducting research on in-hospital resuscitation: the in-hospital 'Utstein style'. A Statement for Health Care Professionals from the American Heart Association, the European Resuscitation Council, the Heart and Stroke Foundation of Canada, the Australian Resuscitation Council, and the Resuscitation Councils of Southern Africa. Circulation. 1997 Apr 15;95(8):2213–39.

Bloom HL et al. Long-term survival after successful inhospital cardiac arrest resuscitation. American *Heart* Journal. 153(5):831–6, 2007 May.

http://www.americanheart.org/presenter.jhtml?identifier=4483. Accessed October 15, 2007.

Case 3
When Coercion Dictates Care

The Patient

Arthur D. is a 78-year-old male with a history of diabetes, hypertension, elevated cholesterol, and remote past tobacco abuse, who presented to the hospital after sustaining three falls at home over the week prior to admission. Arthur's two adult daughters, Brenda and Nancy, accompanied him to the Emergency Department and reported that Arthur had been complaining of dizziness for the last week and they had noticed he had developed an unsteady gait. Although Arthur lived with his elderly wife, Bea, his daughters lived in close proximity to them and saw their parents on a nearly daily basis. Arthur himself offered little in the way of complaints, and often deferred to his daughters to answer questions by the admitting medical residents. His falls resulted in some minor bruising and abrasions, but otherwise he denied any concerning injuries. He denied palpitations, vertigo, light-headedness on standing, headache, blurred vision, or focal weakness. Initial physical and neurological exam was unremarkable, except for some mild gait instability. He was admitted as a 23-Hour Observation Status patient to further evaluate the etiology of his dizziness.

Overnight Arthur was monitored on telemetry, which revealed no noteworthy cardiac events. An orthostatic blood pressure check demonstrated no significant drop in blood pressure or elevation in heart rate with standing. A non-contrast CT scan of his brain showed some nonspecific changes consistent with mild atrophy, but no focal lesions suggesting stroke, mass, or other pathology. A hematological and basic metabolic laboratory work-up was also unrevealing. The Physical Therapy team was consulted to evaluate the patient's gait, and recommended use of a walker as an assist device.

The Ethical Dilemma

After the above evaluation, the medical team planned to discharge the patient with a walker and a referral for outpatient physical therapy. The medical team was paged on rounds by the nurse caring for Arthur, who told the team that Brenda and

From: *Evidence-Based Medical Ethics*
By: J.E. Snyder and C.C. Gauthier © Humana Press, Totowa, NJ

Nancy were in Arthur's room and "very upset" about his planned discharge. They were requesting to talk to the team "immediately" to discuss his case. Arriving in Arthur's room, the team noted his daughters had an angry stance, and wanted to know the results of Arthur's testing to date. After reviewing the findings to date Brenda stated "So, basically, you don't know why my father is falling, and you want to discharge him home?" The senior medical resident replied that the team had ruled out major problems with Arthur's brain and heart that would require prolonged hospitalization, and that further evaluation of his problems could be completed on an outpatient basis with the primary care physician. Nancy stated "There is no way that my father is being discharged before an MRI is done and a neurologist has fully evaluated him." She then added that "I think the hospital CEO would be very interested in knowing that you want to discharge my father without doing this."

Questions for thought and discussion: Do you think that the medical team is obligated to order an MRI and consult a neurologist while Arthur is admitted to the hospital, simply to fulfill the request of his daughter? Does her subsequent threat to involve the hospital CEO change their obligations? Would a threat of litigation?

Question for thought and discussion: Should this particular patient receive further inpatient testing when other patients with similar complaints would undergo their evaluation on an outpatient basis?

To placate Arthur's family a neurology consult was obtained to further evaluate Arthur's gait instability and to determine if an MRI was warranted. Although the neurologist was hesitant to become involved with the case, stating there was "no need" to consult him on the matter, he came to evaluate the patient later that day. His note was brief and offered little additional information to the case. However, he recommended that a brain MRI be performed. The patient, who had now been admitted for more than 24 hours, was changed to Full Admission status and the MRI was ordered. It confirmed the prior findings of the brain CT. The following day, the team reported the results of the neurologist's consultation and MRI to Arthur and his daughters. Nancy demanded that the team consult an otolaryngologist since "maybe my father has an inner ear problem."

Questions for thought and discussion: Should the team continue to comply with the demands of this patient's family? What do you feel are the consequences of not doing so?

Questions for thought and discussion: Does agreeing to the initial demands of a neurology consult and an MRI obligate the medical team to now request inpatient otolaryngology consultation?

The Medicine

Dizziness, vertigo, and gait instability are among the most common complaints of patients seeing a physician, and particularly so in older patients. These complaints are estimated to be present in about 30% or more in those over 65 years old. The yield of radiological tests to determine a diagnosis is generally considered low. In a 2002 case-control study of 211 patients (Colledge et al.), it was shown that the only statistically significant structural abnormality seen in brain/neck MRIs of patients complaining of dizziness is a higher incidence of midbrain white matter lesions. Overall, routine MRI was considered by the study investigators not to be a useful modality for determining etiology of this complaint.

The Law

Although there are no specific federal or state laws to address the ethical issues at hand in this case, there are rules and regulations pertaining to billing for hospital services that must be considered. This patient is 78–years-old and the costs of his initial hospitalization are likely covered by Medicare Part A, which is insurance coverage for medical needs such as inpatient hospitalization, stay at skilled nursing facilities (SNFs), hospice care, and some other situations. Medicare Part A is a system that generally covers patients over 65 years of age, patients with end-stage renal disease requiring hemodialysis or kidney transplantation, and patients with certain disabilities. Medicare Part B, which is insurance coverage for outpatient services, also includes coverage for services and supplies such as those associated with this patient's physical therapy needs.

When a patient is admitted to the hospital, the hospital's utilization review committee will evaluate the patient's level of care based on specific and well-established criteria. By this, the need for initial evaluation and treatment on an inpatient basis, based on medical necessity, must be determined and documented in the patient chart. If the patient does not meet the requirements for inpatient admission, or in other words the entire hospitalization could have been avoided and the patient care completed entirely on an outpatient basis, then hospitalization is considered "Medically Unnecessary Services" and there will likely be no Medicare reimbursement for the hospitalization.

If a patient has, in fact, met criteria for hospitalization, then Medicare will reimburse the hospital for appropriate care. Once the patient has been medically stabilized and is ready for discharge, there will be no reimbursement for further inpatient services. At this point, if a safe discharge plan has been made for the patient and they refuse to accept it, the hospital can give the patient a letter telling them that they may be personally responsible for all costs of their care from that point forward. The patient can appeal this decision with Medicare, but only if they win their appeal will they not be responsible for these medical expenses.

In this case, as soon as the primary physician caring for Arthur determines that the medically necessary inpatient observation period – including diagnostic evaluation and medical intervention – is completed, there will be no reimbursement for further inpatient testing, evaluation, and care. If Arthur remains hospitalized beyond this period, he can be presented with a letter stating that all medical expenses related to his hospital stay from that point forward may become his fiscal responsibility. If he appeals this decision and loses, he can potentially receive a bill for thousands of dollars' worth of medical services rendered. If he is unable to pay for these services, the hospital will assume a financial loss for his medical care. Hospitals can acquire significant debts from such reimbursement denials. Public hospitals, in particular, suffer significant financial pressures by these accumulating debts, in addition to shortfalls from the unpaid medical bills of uninsured patients.

The Ethics

Based on the *Principle of Veracity*, Arthur should have been given complete information about the results of the tests performed during the observation period, the conclusions drawn from these, and the medical team's recommendations for a walker and outpatient physical therapy. If Arthur had agreed, his daughters could have been given the same information. In this case Arthur and his family seem to have been told he was being discharged, without being given this information. It was not until the daughters confronted the medical team that they received the results of their father's tests. They may have felt that they were being dismissed by the medical team without being given this important information. This may have contributed to their feelings of anger and their demands for further tests.

The daughters' concern about their father's recent falls and the fear that he would suffer more serious consequences if he returned home and fell again may also have contributed to their hostility and demanding attitude. The medical team might have avoided this confrontation with the family by making sure that they were all well-informed about the tests results and why Arthur was being discharged. They could have reassured Arthur and his daughters about the future with their recommendation for a walker and the referral to physical therapy.

However, the medical team is not obligated to order an MRI, a neurology consult, or full hospital admission for Arthur. The only justification for these actions would be based on the *Principles of Beneficence* and *Non-Maleficence*, if the medical team believed these tests would benefit the patient and that the benefits outweighed the possible risks. Hospital admission, for example, exposes patients to hospital-borne infections, so there must be a good reason to admit a patient to the hospital. In this case the medical team did not believe these interventions would be in the best interests of their patient.

Moreover, neither the patient nor the family has the moral right to demand specific medical interventions. The *Principle of Respect for Autonomy* requires that capable patients be permitted to accept or refuse treatment alternatives recommended by their physicians. This does not mean that patients or families can demand whatever procedures or consults they want.

When the daughters demanded an MRI and a neurologist's evaluation for their father, the senior medical resident should have explained why these steps were not warranted in their father's case. This may have helped them to better understand and accept the judgment of the medical team. Arthur and his family could also have been told that the costs of further testing and extended hospitalization would be their responsibility and may not be covered by Medicare.

Whatever the outcome of these efforts to communicate rationally with Arthur's daughters, no threat could change the actual obligation of the medical team toward their patient, although they may be concerned about involving the hospital administration and facing a lawsuit. In response to situations like this, physicians should take responsibility for practicing medicine according to their training, experience, and medical judgment. Physicians can preserve their own dignity by resisting the temptation to order tests and procedures simply because patients or families demand them and back up those demands with threats, like that of litigation. If the medical team members agree on what procedures the patient needs and what procedures are not necessary, they should be confident that they can justify their judgment to the administration and in a court of law.

Another concern is the fairness of ordering further inpatient testing for Arthur, based on his daughters' threats, when other patients with similar conditions undergo their evaluations on an outpatient basis. The *Principle of Distributive Justice* requires that health care resources be distributed in a fair way. Using this principle the medical team might think about why the hospital resources needed for inpatient testing should be used for this patient, while they will not be used for other similar patients. Two good reasons for providing an individual patient with health care resources are that the patient needs them and that they are the best use of those resources. An example is the system in place for allocation of donated organs to persons in need who are deemed to be suitable candidates for this limited resource. Threats are not a good reason to provide resources to one patient and not to another.

The medical team should not continue to comply with the demands of this patient's family, once the neurology consult and MRI have been done. The daughters should be told to ask their father's primary care physician to refer him to an inner ear specialist once he has been discharged. What are the possible consequences of this? The daughters may become so angry that they seek care elsewhere, which might be better for all concerned. They may threaten to involve the hospital administration, but this may work to the advantage of the medical team. The administration is likely to support the medical team if they are in agreement on their decisions and confident in their judgments. The threat of a lawsuit is not really viable in this case. It is doubtful that any reputable lawyer would take the case since the patient has been given an extensive battery of tests and has not been harmed in any way.

Agreeing to the initial demands of the MRI and the neurology consult does not obligate the medical team to request inpatient otolaryngology consultation.

Although the medical team had no moral obligation to order an MRI, a neurology consult, or admit the patient to the hospital, they may have done so to assure the family that everything possible had been tried to further diagnose the patient's problem before discharge. They have done as much as they are morally required to do and they should refuse this latest demand to preserve their own dignity and authority as medical practitioners.

At this point, the medical team should offer to send a copy of Arthur's medical record, including test results and consults, to his primary care physician, and provide Arthur with a referral for outpatient physical therapy, as planned. They might also advise the family to make an appointment with Arthur's primary care physician to pursue any further tests and consults that may be warranted, on an outpatient basis.

The Formulation

Now that the evidence-based medicine, legal precedent, and relevant ethical principles for this case have been reviewed, formulate a strategy to address the ethical conflicts in this case. If necessary, perform additional research into local and state laws and hospital regulations. Consider delving further into the background medical literature to assist with making sound therapeutic decisions. Devise a treatment approach that addresses the needs of the patient and his family, that is both ethically and medically sound, and that is culturally competent. Ensure that the strategy employs fair and appropriate utilization of medical resources, and that the approach is practical and feasible within the limits of the medical system at large. Work out a clear and professional way to communicate the proposal to the patient and his family. Attempt to foresee challenges that may arise in conveying or implementing the plan. Determine what follow-up will be necessary to ensure that the chosen strategy remains successful for the patient in the long-term. Reflect on how the knowledge and skills learned from this case can be used to improve the care of patients that may be encountered in future practice.

Afterthoughts

In this case the medical team caring for Arthur felt coerced into prolonging the hospitalization of the patient, as well as ordering inpatient consults and testing that they did not feel were medically necessary. A fear of reprimand from the hospital administration, perhaps combined with a fear of litigation, led this team to question their initial best judgment and alter their course of care. How would this case have been different if Arthur himself were making the demands, as opposed to just being complicit with his daughters' demands? How much do you think fear of litigation drives physicians to go against their better judgment? Do you think that this contributes significantly to the rising national debt that is due to medical expenses?

Annotated References/Further Information

http://www.cms.hhs.gov. Accessed October 15, 2007.

Code of Medical Ethics of the American Medical Association: Current Opinions with Annotations, 2006–2007 Edition. Council on Ethical and Judicial Affairs. Annotations prepared by the Southern Illinois University Schools of Medicine and Law.

Snyder L, JD and Leffler C, JD. Ethics Manual, Fifth Edition. Ethics and Human Rights Committee, American College of Physicians. Annals of Internal Medicine. 142(7):560–582, 5 April 2005.

Colledge N. et al. Magnetic resonance brain imaging in people with dizziness: a comparison with non-dizzy people. Journal of Neurology, Neurosurgery & Psychiatry. 72(5):587–9, 2002 May.

Case 4
When a Spouse Is Estranged

The Patient

Sharon M. is a 49-year-old female who was diagnosed with Stage IIA cancer of the left breast three years prior to admission. At the time of diagnosis she opted to undergo surgical lumpectomy, but declined postoperative radiation therapy. As she was pre-menopausal at the time and the tumor was estrogen receptor-positive, she was started on tamoxifen therapy and followed thereafter with regular clinical examinations and surveillance mammography. In the ensuing months Sharon felt well and had no significant medical problems.

Sharon came to the Emergency Department because her best friend, Jane, encouraged her to do so. At the time of admission Sharon presented with complaints of progressive shortness of breath over two months' time. She denied chest pain, palpitations, lower extremity swelling, fevers, chills, or diaphoresis. She did note that her appetite had recently been poor, and that she had an unintentional 20-pound weight loss over the past few months. Additionally, she had a mild persistent headache for the past several weeks. On physical exam she was noted by auscultation to have absent breath sounds at the base of her left lung. On neurological examination there was a slight deficit in her mental status examination and the admitting medical team also noted a mild facial asymmetry. Chest radiography demonstrated a large left-sided pleural effusion and several soft tissue masses consistent with metastatic cancer. CT of the brain demonstrated several small densities consistent with metastatic lesions, with mild edema and a slight mass effect. An MRI of the brain additionally showed leptomeningeal involvement. She was started on supplemental oxygen and morphine for her breathing discomfort, dexamethasone for the brain metastases, and a radiation oncology consultation was obtained.

The Ethical Dilemma

Unfortunately, as Sharon began to undergo whole brain radiation therapy for treatment of her metastases, she became increasingly confused, somnolent, and minimally verbal. Her friend, Jane, vigilantly remained at her bedside to offer her comfort and often

From: *Evidence-Based Medical Ethics*
By: J.E. Snyder and C.C. Gauthier © Humana Press, Totowa, NJ

assisted Sharon's nurses with her bathing and other care. Jane told the medical team that Sharon had been married at one time to a physically- and verbally-abusive alcoholic man named Walter, and that she had left him about 10 years prior to admission. Although they had not ever been formally divorced, Sharon had clearly stated to Jane on a number of occasions in the past that she did not want Walter to "know anything about me or my health." Jane also believed that if Sharon could say so herself, she would most likely only want her medical care to focus on comfort and noninvasive measures, given the futility of treating her underlying illness. During a brief, but more lucid, moment in her care Sharon confirmed both these facts by nodding yes to the medical team.

The hospice team was consulted to assist with Sharon's care and for possible transfer to an inpatient hospice facility. Although willing to accept Sharon to the facility based on her medical needs and purported wishes, Sharon had become unable to give verbal or written consent to the transfer. She had not pre-arranged to legally have Jane or anyone else become her health care agent prior to the current illness, and had never filled out a Living Will. Sharon's parents were no longer alive and she didn't have children. The hospice team stated that only Sharon's husband could give consent to admit her to the inpatient hospice facility since they were technically still married.

Questions for thought and discussion: Who can give consent for Sharon in this situation? Should the medical team contact Sharon's husband, Walter, even if she expressed a strong desire to exclude him from her life and medical care?

Question for thought and discussion: Would this situation change if Sharon still had living parents or if she had adult children?

The Medicine

According to the Centers for Disease Control and Prevention (CDC), breast cancer is the most common form of cancer in women aside from non-melanoma skin cancer, and the second most common cause of cancer death. In 2003, there were 181,646 women diagnosed with breast cancer in the United States and 41,619 deaths from the disease. In other terms, a woman dies from breast cancer every 13 minutes in the U.S. The best chance for survival from breast cancer is early detection and treatment. When breast cancer recurs, treatment is rarely curative. In stage IV disease, where the cancer has spread to other organs, treatment is palliative only, with the goal of maximizing quality and duration of life.

The Law

If the medical team believes Sharon was capable of making her own medical decisions during her lucid periods, then her wishes not to involve her estranged husband and to receive only comfort care, expressed nonverbally, should be written in her medical chart and honored.

Medical confidentiality is legally protected in most states. Contacting this patient's husband would definitely violate medical confidentiality. The husband would need to be given detailed information about Sharon's medical condition, her diagnosis and her prognosis if he were the correct person to give informed consent for her admission to hospice. However, the patient's specific desire that her husband not be given this information further strengthens the legal protection of her medical privacy.

Under the federal Health Insurance Portability and Accountability Act (HIPAA) Privacy Rule, "individually identifiable health information" may not be released by health care providers without the patient's consent. There are a number of exceptions to this rule, including release to a friend or family member involved in the patient's care (unless the patients objects), or when health care providers believe the release of information would be in the patient's best interests.

The medical team is legally authorized to limit Sharon's care to comfort measures, based on her own wishes expressed in response to their questions. However, Sharon was not asked specifically about transfer to the hospice facility. In most states, if a patient has not appointed a health care agent, the patient's spouse would become her surrogate decision maker. This may be the basis for the hospice team's statement that only Sharon's husband can give consent for her admission to hospice. In this case, however, Sharon's express wishes not to involve her estranged husband rule him out as her surrogate decision maker.

Legally, no one is presently authorized to give consent for this transfer and Sharon may have to receive end-of-life comfort care in the hospital. One alternative solution would be for the medical team to approach the hospital attorney about asking a court to appoint Jane as Sharon's legal guardian, if Jane is willing to take on the responsibility of making medical decisions for her friend.

If Sharon had living parents or adult children, the situation would be very different. In most states, parents and children follow the spouse in priority as surrogate decision makers. Once Sharon had refused to have her husband involved, one of her parents or one of her adult children would have become the legally-authorized decision maker and would have been able to give consent for her transfer to the inpatient hospice facility.

The Ethics

Since the medical team has consulted the hospice care team, it is clear that they believe comfort care would be best for Sharon at this stage of her illness. They will have compared the harmful side effects and possible benefits of continued aggressive treatment, using the *Principles of Beneficence* and *Non-Maleficence*. They will have decided that such treatment would not be effective in meeting any reasonable goals for their patient and would not be in her best interests.

The medical team may have shared this information with Jane, since she seems to understand that further aggressive treatment would be futile. In that case, it could be argued that the medical team has violated Sharon's confidentiality.

On the other hand, Sharon obviously trusts her friend, since she came to the Emergency Department because Jane encouraged her to do so. Jane is clearly concerned about Sharon and has remained in the hospital with her and helped with her personal care. It was also very important that the medical team communicate honestly with Jane to learn as much as possible about Sharon, including her family situation and her wishes regarding medical treatment at the end of life. Finally, being at Sharon's bedside and observing her deterioration, Jane may have independently come to the conclusion that continued cancer treatment for Sharon would be futile.

Based on the *Principle of Respect for Autonomy* and the *Principle of Respect for Dignity*, the medical team should not contact Sharon's estranged husband. Sharon has expressed her desire that he not be involved in her medical care. Her wishes on this issue should be respected. Moreover, to contact Walter against her wishes would violate Sharon's privacy, an important aspect of human dignity.

It appears that no one is able to give consent for Sharon to be transferred to the inpatient hospice facility, unless a guardian is appointed for her by the court. However, the medical team knows what her wishes are: she wants to be given comfort care instead of aggressive treatment. They may also believe that this sort of care is best provided in a hospice facility. If so, they may be able to convince the hospice team, using the *Principle of Beneficence* and the *Principle of Respect for Autonomy*, to accept the transfer based on Sharon's best interests and her own desires.

Ethically, a parent or adult child would be the natural choice to be Sharon's surrogate decision maker, once she had expressly excluded her estranged husband. Ideally, these close family members would have Sharon's best interests in mind and would want her to receive the care she wanted in the best possible setting. Of course, whoever served as her decision maker would need to be fully informed about her diagnosis and prognosis, and the reasons the medical team has recommended transfer to the inpatient hospice facility.

The Formulation

Now that the evidence-based medicine, legal precedent, and relevant ethical principles for this case have been reviewed, formulate a strategy to address the ethical conflicts in this case. If necessary, perform additional research into local and state laws and hospital regulations. Consider delving further into the background medical literature to assist with making sound therapeutic decisions. Devise a treatment approach that addresses the needs of the patient, her friend Jane, and her family, that is both ethically and medically sound, and that is culturally competent. Ensure that the strategy employs fair and appropriate utilization of medical resources, and that the approach is practical and feasible within the limits of the medical system at large. Work out a clear and professional way to communicate the proposal to the patient, her friend Jane, and (if necessary) Walter. Attempt to foresee challenges that may arise in conveying or implementing the plan. Determine what follow-up will be necessary to ensure that the chosen strategy remains successful for the patient in the long-term. Reflect on how the knowledge and skills learned from this case can be used to improve the care of patients that may be encountered in future practice.

Afterthoughts

In this case, Sharon had not completed a Living Will or Health Care Power of Attorney and legally had no one, but her estranged husband, to legally speak on her behalf when she became incapacitated. Ultimately it was clear what she would have wanted to have done, but it was an an ethical challenge to obtain this for her. It is not uncommon for patients to have no official documentation of their wishes in such situations. How would this case have been different if the medical team had determined that Sharon's husband Walter had been sober for five years and felt repentant about his past abusive relationship with Sharon? What if they found out that Walter had passed away two years previously – who would have been Sharon's surrogate decision maker then? Would the case have been different if Sharon and Walter were still together, but the team found out that he was an active alcoholic and physically abusive to Sharon – would he be qualified to speak for her in this case?

Annotated References/Further Information

Code of Medical Ethics of the American Medical Association: Current Opinions with Annotations, 2006–2007 Edition. Council on Ethical and Judicial Affairs. Annotations prepared by the Southern Illinois University Schools of Medicine and Law. The subsections on Medical Futility in End-of-Life Care (2.037), Confidentiality (5.05), and Surrogate Decision Making (8.081) are relevant to this case.

Snyder L, JD, and Leffler C, JD. Ethics Manual, Fifth Edition. Ethics and Human Rights Committee, American College of Physicians. Annals of Internal Medicine. 142(7):560–582, 5 April 2005. The subsections on Confidentiality, Informed Consent, Patients Near the End of Life, and Making Decisions Near the End of Life address a number of issues raised by this case.

Health Insurance Portability and Accountability Act Privacy Rule 45 CFR 164.

Subsection 164.510 explains the HIPAA Privacy Rule as it applies to health care providers and their patients.

http://www.cdc.gov/cancer/breast/statistics/. Accessed on October 15, 2007.

http://www.cancer.gov/cancertopics/pdq/treatment/breast/HealthProfessional/page8. Accessed on October 15, 2007.

Case 5
When a Patient Is Behaving Badly

The Patient

Jeffrey T. is a 25-year-old male with an extensive history of ongoing intravenous heroin and cocaine use, a bioprosthetic tricuspid valve replacement for infective endocarditis three years prior to admission, and subsequent prosthetic valve infective endocarditis one year prior to admission that was managed medically with antibiotic therapy. He presented to the hospital with fevers, chest pain, shortness of breath, and the chief complaint, "I think I have endocarditis again." Review of his medical record found documentation of past hospitalizations for two drug overdoses, a diagnosis of substance-induced mood disorder, and a "likely" diagnosis of antisocial personality disorder. In the Emergency Department, Jeffrey's physical exam was notable for a temperature of 103.2°F and a loud murmur on cardiac auscultation. There were no obvious peripheral manifestations of endocarditis on physical exam. An electrocardiogram was within normal limits. Chest radiography demonstrated two small, nodular lesions in the right middle lobe concerning for septic emboli. A urine toxicology screen was positive for cocaine and opiates.

Jeffrey was admitted to the telemetry floor for work-up of his fevers and likely recurrent endocarditis. Three sets of blood cultures were drawn per protocol and empiric antibiotic therapy for infective endocarditis was started due to high suspicion for this diagnosis. A transthoracic echocardiogram showed no obvious valvular vegetations. The cardiology team was consulted to assist with the management of his case, and to perform a transesophageal echocardiogram the next morning.

The Ethical Dilemma

On the evening of admission, the medical team was paged several times by Jeffrey's nurse as he was demanding analgesic medication for his chest pain beyond what was originally ordered by the medical team. Once pain medication was given, he would leave his hospital room for up to 90 minutes at a time without telling his nurse where he was going or when he would return. His absences from his hospital room made it difficult for the nurse to administer the empiric antibiotics ordered by the medical

From: *Evidence-Based Medical Ethics*
By: J.E. Snyder and C.C. Gauthier © Humana Press, Totowa, NJ

team. At one point in the middle of the night, hospital security was called to Jeffrey's room as he had six visitors in his room, past hospital visiting hours, who were loud and disturbing other patients. Additionally, Jeffrey's nurse was concerned that drug deals were being made in the hospital room at that time. The security guards escorted the visitors from the hospital. Two hours later Jeffrey's nurse found him in his hospital bed, poorly responsive and barely breathing. The medical team quickly responded to the patient and administered intravenous narcan with good response. Jeffrey's nurse complained to the medical team that Jeffrey had been keeping her so busy that she barely had time to care for her other patients. A nursing assistant was assigned to sit in Jeffrey's room and closely observe him the rest of the evening.

Question for thought and discussion: If Jeffrey has a fatal drug overdose while on hospital premises, are the hospital and the medical team caring for him liable for legal action against them?

The next morning the team approached Jeffrey on rounds to obtain his verbal agreement to comply with his caregivers' needs to diagnose and treat his condition, and to be less disruptive to other patients. He agreed and expressed remorse for his behavior overnight. However, an hour later, when the nursing assistant observing him momentarily had stepped out of his room, Jeffrey left the hospital floor for over an hour and missed the assigned time for his transesophageal echocardiogram.

Question for thought and discussion: How should the medical team react to Jeffrey's continued behavior?

Questions for thought and discussion: Would a written behavioral contract with Jeffrey, with explicit rules for him to follow, be of use? If so, can the medical team discharge him from the hospital if he breaks the rules in such a contract, knowing that he could potentially die if he were not treated for his medical condition? Are such contracts ethical, and do they stand up to legal scrutiny?

Questions for thought and discussion: Is confining Jeffrey to his hospital room and denying him visitors a violation of his rights? Is assigning someone to constantly observe Jeffrey in his hospital room a violation of his privacy?

Question for thought and discussion: Do you think Jeffrey can be involuntarily committed to the hospital by the psychiatry team if he is deemed unsafe to himself or others?

The Medicine

Among cases of infective endocarditis, 10 to 15 percent involve prosthetic heart valves. Due to small vegetation sizes and postoperative changes, transesophageal echocardiography is often needed to make the diagnosis. Mortality rates can be high (40–50%) due to the frequency of complications, the long duration of antibiotic treatment needed for cure, and technically difficult surgeries required in some cases. Intravenous drug users have a higher incidence of recurrent disease due to continued injection behaviors, and have higher mortality rates from endocarditis than non-drug users. Ongoing substance abuse and poor adherence with recommendations may prolong length of hospital stays for these patients, and may make cardiac surgeons less eager to operatively intervene on complicated cases.

The Law

The hospital at large may have been liable if Jeffrey had died from a drug overdose while admitted as a patient there. It is doubtful, however, that the medical team themselves would also be liable, since they would not be continually present on the patient's floor and could not be held responsible for his behavior in the hospital.

Hospital discharge as a potential consequence for noncompliance with the behavioral contract would be legally risky in this case. The patient has sought medical treatment and should not be discharged, against his will, with a life-threatening condition. If the hospital had ever received federal money for construction or renovation under the Public Health Service Act (1975), it must make services available to all those who reside in the service area, without discrimination on any basis unrelated to the patient's need or the availability of the service. Discharging Jeffrey without making plans for his continued care may also be considered a form of patient abandonment. On the other hand, Jeffrey's actions (leaving his room and missing scheduled tests and antibiotic therapy) could be interpreted as refusal of recommended treatment.

Confining Jeffrey to his room may also violate the Conditions of Participation for hospitals receiving Medicare and Medicaid funds, established by the federal Health Care Financing Administration. Under a list of Patients' Rights, these conditions include a standard on restraint or seclusion. Seclusion is defined as "involuntary confinement of a patient alone in a room or area from which the patient is physically prevented from leaving" (42 Code of Federal Regulations Part 482.13 (e)). Seclusion "may only be used for the management of violent or self-destructive behavior" (42 Code of Federal Regulations Part 482.13 (e)). The medical team needs to decide if Jeffrey's behavior meets this standard. They should also consult hospital policies on confinement before including this as a consequence in a written behavioral contract.

The Ethics

Jeffrey cannot be permitted to continue with this behavior. For his own safety and that of the hospital staff and other patients, he should not be allowed to leave his room for long periods of time. It is particularly worrisome that Jeffrey is wandering around the hospital unsupervised and may pose a threat to other patients, particularly if he is under the influence of illegal drugs.

Based on the *Principle of Beneficence*, the medical team should consider the good to be promoted for this patient and the harm to be prevented. The patient was admitted to the hospital for further diagnostic tests and antibiotic therapy for a possible life-threatening condition. By leaving his room and missing scheduled tests and treatments Jeffrey is preventing the hospital staff from accomplishing these purposes and jeopardizing his own health.

Based on the *Principle of Distributive Justice*, the medical team must also consider the effects of Jeffrey's behavior on other patients, including the disturbance caused by his unruly visitors. Dealing with Jeffrey is also taking time away from the care of other patients on his floor. The care provided to one patient in the hospital should not threaten the care that other patients receive, without some convincing justification.

The *Principle of Respect for Autonomy* would allow Jeffrey to make his own decisions about what tests and treatments to accept, and allow him to refuse these as well. However, once he has agreed to the hospital admission and consented to the recommended diagnostic tests and antibiotic therapy, he should not be permitted to waste hospital resources by missing his scheduled tests and treatments. In this case the fair distribution of medical resources may place a limit on respect for the patient's autonomy.

The medical team must take charge of the situation by setting limits on Jeffrey's actions while he is a patient in the hospital. They should also meet with the nursing staff caring for Jeffrey so that they also understand these limits. Now that a verbal contract has proved ineffective, a written behavioral contract may be a good option.

Before Jeffrey is asked to sign a written contract, a psychiatric consult might be ordered to confirm that he is capable of understanding the consequences of noncompliance with the contract. Also, someone on the medical team should have a conversation with Jeffrey to establish goals for his medical treatment that he can accept and to which he agrees. Based on the *Principle of Respect for Dignity*, this conversation should begin with Jeffrey's own goals for treatment and may also include inquiries about family relationships that could provide him with support during his hospital stay. It will also be important to confirm that he understands the risks to his life and health if he avoids the recommended tests and therapy.

A written behavioral contract could include these agreed-upon treatment goals and rules for Jeffrey's behavior. It should also include consequences of noncompliance, such as limits on when he can leave his room and where he can go in the hospital, and restricting his visitors to responsible adult family members. These actions would not constitute "seclusion" and would not violate any of Jeffrey's rights, within the hospital context. All rights have limits, based on preventing harm to others and, in some cases, preventing harm to the individual concerned as well. Restricting his movements within the hospital and limiting those who visit him may be necessary to protect other patients from being disturbed by Jeffrey or his visitors.

Limiting his visitors may also be necessary to keep Jeffrey from obtaining illegal drugs. Controlling the times he can leave his room may be necessary for the diagnosis and treatment of his condition, averting the harm of further damage to his heart and possible death. Interference with Jeffrey's choices and actions for his own good would be an example of paternalism. However, this may be justified by the *Principle of Beneficence*, similar to the use of restraints to prevent a patient from pulling out an IV line or feeding tube. The use of 24-hour patient monitoring with a nursing assistant as a consequence of Jeffrey's noncompliance might be considered to be an unnecessary violation of privacy, particularly if Jeffrey's visitors and his movements within the hospital are already restricted.

From an ethical perspective, discharge from the hospital should not be written into the behavioral contract as a consequence of noncompliance. Jeffrey should not be discharged, against his will, without arrangements for the continuity of his care. His condition may be life-threatening, and without further tests and antibiotic therapy his life may be in danger. However, Jeffrey must understand that his continued refusal of tests and treatments makes it difficult to justify keeping him in the hospital.

As a last resort, the medical team could seek involuntary commitment if a written contract proves unsuccessful in controlling Jeffrey's behavior. Short-term commitment, for evaluative purposes, may be possible. However, long-term commitment is unlikely, as long as Jeffrey is determined by a judge to be legally competent.

The Formulation

Now that the evidence-based medicine, legal precedent, and relevant ethical principles for this case have been reviewed, formulate a strategy to address the ethical conflicts in this case. If necessary, perform additional research into local and state laws and hospital regulations. Consider delving further into the background medical literature to assist with making sound therapeutic decisions. Devise a treatment approach that addresses the needs of the patient and his family, that is both ethically and medically sound, and that is culturally competent. Ensure that the strategy employs fair and appropriate utilization of medical resources, and that the approach is practical and feasible within the limits of the medical system at large. Work out a clear and professional way to communicate the proposal to the patient and his family. Attempt to foresee challenges that may arise in conveying or implementing the plan. Determine what follow-up will be necessary to ensure that the chosen strategy remains successful for the patient in the long-term. Reflect on how the knowledge and skills learned from this case can be used to improve the care of patients that may be encountered in future practice.

Afterthoughts

Jeffrey's behavior and possible illegal activity in the hospital limited his medical team's ability to provide appropriate care. Additionally, he had a near fatal drug overdose on hospital premises, creating a liability concern. How would this case have been different if Jeffrey's behavior included angry outbursts and aggressiveness toward his caregivers and their personal safety was threatened? At what point can health care providers disavow themselves of caring for a difficult patient?

Annotated References/Further Information

Code of Medical Ethics of the American Medical Association: Current Opinions with Annotations, 2006–2007 Edition. Council on Ethical and Judicial Affairs. Annotations prepared by the Southern Illinois University Schools of Medicine and Law. The subsection on Termination of the Physician-Patient Relationship (8.115) is relevant to this case.

Snyder L, JD and Leffler C, JD. Ethics Manual, Fifth Edition. Ethics and Human Rights Committee, American College of Physicians. Annals of Internal Medicine. 142(7):560–582, 5 April 2005. The subsection on Initiating and Discontinuing the Patient-Physician Relationship is particularly relevant to the issue of patient abandonment raised in this case.

Medicare and Medicaid Programs; Hospital Conditions of Participation: Patients' Rights; Final Rule (2006). 42 Code of Federal Regulations Part 482. 482.13 (e) Standard: Restraint or seclusion.

www.cms.hhs.gov/CFCsAndCoPs/downloads/finalpatientrightsrule.pdf. Accessed October 15, 2007.

Beauchamp TL and Childress JF. Principles of Biomedical Ethics, Fifth Edition, Oxford University Press, 2001. See pages 176–191 for a good discussion of medical paternalism.

Public Health Service Act (1975) Title VI. 42 Code of Federal Regulations Part 124.603.

www.hhs.gov/ocr/hbcsreg.html. Accessed October 15, 2007.

Beynon RP et al. Infective endocarditis. BMJ. 333(7563):334–9, 2006 Aug 12.

Case 6
When a Patient Becomes Agitated

The Patient

Kathleen W. is an 82-year-old female with a history of hypertension, elevated cholesterol, diabetes mellitus, past myocardial infarction, and congestive heart failure. She presented to the Emergency Department with a two-week history of progressively worsening shortness of breath, a 10 pound weight gain during this period and worsening lower extremity edema. She denied chest pain, palpitations, fevers, or any other complaints. On further review the admitting medical team determined that Kathleen could provide only an incomplete list of her home medications and did not know any of the correct dosages. Additionally, she did not follow a low-sodium diet as is recommended to all patients with congestive heart failure. The medical team confirmed her medication information with her pharmacy and the pharmacist stated that Kathleen had not refilled some of her medications in the preceding two months.

On physical exam, Kathleen was found to have mild hypoxia, a rapid and irregular heart rate, diffuse râles on pulmonary examination, and bilateral lower extremity pitting edema to the knees. A hematological and basic metabolic laboratory work-up was unrevealing. A thyroid-stimulating hormone (TSH) level was within normal limits. An electrocardiogram demonstrated a new diagnosis of atrial fibrillation with a rapid ventricular response. The patient was started on a diltiazem drip for heart rate control, with subsequent stabilization. Chest radiography confirmed a diagnosis of congestive heart failure, and the patient was started on diuretics, an ACE-inhibitor, and supplemental oxygen. She was admitted to the cardiac telemetry floor for further evaluation and management.

The Ethical Dilemma

At 2:00 AM on the night of admission Kathleen's nurse found that she had pulled out her IV. This was the line that her diltiazem drip was running through, and the former IV site was bleeding considerably. Kathleen was notably agitated and

From: *Evidence-Based Medical Ethics*
By: J.E. Snyder and C.C. Gauthier © Humana Press, Totowa, NJ

somewhat combative with the nurse as she attempted to dress the wound. The medical team was called to evaluate her agitation and determined that it was most likely due to delirium or "sundowning." They prescribed a benzodiazepine for mild sedation. However, Kathleen's confusion and agitation worsened, likely from a paradoxical reaction to the benzodiazepine. Haloperidol was then administered with little effect, although a second, higher strength dose was more successful in calming her. Soft wrist restraints were placed to control Kathleen's arm movements so the nurse could place a new peripheral IV and restart the diltiazem drip, and to prevent Kathleen from pulling the new IV out.

When the medical team rounded on Kathleen the next morning, she was sleeping soundly and barely arousable to voice due to residual effect of the benzodiazepine and haloperidol. Kathleen's son, Thomas, was in his mother's room and demanded, "why is my mother sedated and tied to the bed?" The team explained to Thomas the concept of "sundowning," and how it was necessary to calm his mother's agitation to adequately treat her overnight. Thomas remained perturbed despite the explanation and asked that the wrist restraints be removed now since his mother was highly sedated. He further stated that no physical or chemical restraints should be used in the future without his express consent.

Questions for thought and discussion: Is the use of chemical restraints (sedation) ethical in this case? What about the use of physical restraints?

Question for thought and discussion: How could the medical team have managed Kathleen's case differently to prevent some of Thomas's frustration?

Question for thought and discussion: Can the medical team continue to use chemical or physical restraints in caring for Kathleen if Thomas does not give his permission for them to do so?

The Medicine

In a systematic review of delirium superimposed on dementia (Fick et al.), it was estimated to complicate at least 24% of hospitalizations of persons living with dementia. This phenomenon not only affects the quality of care an individual patient receives in the hospital, it impacts patient mortality and morbidity and carries a significant financial burden related to increased hospital lengths of stay. As common as delirium may occur in this subpopulation, it is always essential to formally evaluate a patient with underlying dementia who develops acute mental status changes and not simply label them as "sundowning." In general, treatment of delirium in the person with dementia primarily involves removal of contributing medications or other factors, in combination with routine supportive care. For patients at high risk for self-harm, the use of psychotropic medications and physical restraints may be necessary, although the latter may increase agitation or cause the patient injury.

The Law

The federal Health Care Financing Administration has established standards for the use of chemical and physical restraints as part of the Conditions of Participation for hospitals that receive Medicare or Medicaid funds. According to these standards, restraints, whether physical or chemical, "may only be imposed to ensure the immediate physical safety of the patient, a staff member, or others and must be discontinued at the earliest possible time" (42 Code of Federal Regulations Part 482.13 (e)). Before restraints are used, less restrictive interventions must be shown to be ineffective to protect the patient, staff, or others from harm. Use of restraints must be ordered by the attending physician and the order may only be renewed for four hours at a time, up to a total of 24 hours. The physician must see and evaluate the patient one hour after initiating restraints. At the end of the 24-hour period the attending physician must see and assess the patient before a new order may be written. Individual hospitals often have additional regulations in place to ensure that a patient's rights are not violated and that safety is maintained in the setting of restraint use.

Use of chemical restraints for Kathleen was necessary for her immediate safety, since she had pulled out the IV that was delivering medication to control her atrial fibrillation. Without this medication, her condition would have worsened and may have in theory resulted in death. Using soft wrist straps as physical restraints may also be justified by the need to place a new IV and to keep her from pulling it out.

If the medical team has followed the other guidelines for the use of restraints, there should not be a problem with the federal law. However, it might be questioned why the wrist restraints were not removed once she was adequately sedated. A member of the medical team should have evaluated Kathleen one hour after initiating sedation and physical restraints. The restraint orders should also have been renewed, if still needed, four hours after they were first written.

The medical team should ensure they are complying with relevant hospital policies and any applicable state laws in their use of physical and chemical restraints. Both hospital policies and state laws may vary and may be more restrictive than the federal requirements.

The Ethics

Based on the *Principle of Beneficence*, chemical restraints could be justified as necessary in preventing harm to Kathleen. Pulling out another IV would have caused further injury to the site and loss of blood. Kathleen also needs the prescribed medications to regulate her heart rate and treat her congestive heart failure. Without them her condition could become life-threatening. The medical team will have considered the sedatives' side effects, using the *Principle of Non-Maleficence*, and compared this harm to the harm that would be prevented if Kathleen was able to receive her cardiac medications. It appears that the medical team decided Kathleen would suffer more harm if she was not sedated.

The use of physical restraints might be justified in the same way. On the other hand, one might question the further need for wrist straps once Kathleen calmed down from the effects of the sedatives. However, the medical team may have been uncertain whether she would remain calm for the rest of the night.

Considering the *Principle of Respect for Autonomy*, Kathleen's capacity for decision making was not questioned by the medical team in the Emergency Department. We can assume that she agreed to the hospital admission and the medications that were recommended. Once Kathleen began to experience delirium, her capacity would become questionable. Most likely she was not even consciously aware of pulling out her IV. As her agitation and confusion increased, Kathleen was clearly not capable of making her own decisions. However, the 2:00 AM decision to use chemical and physical restraints could be understood as an emergency decision, with no time to ask a surrogate decision maker to grant permission.

At the time of her admission to the hospital, the medical team or another hospital representative would have inquired about Kathleen's family members and whether or not she had a Living Will or Power of Attorney for Health Care. If Kathleen does not have an advance directive, if she is a widow, and if Thomas is her only child, then he would automatically become her surrogate decision maker in most states.

With this in mind, it could be argued that one of the physicians on the medical team should have called Thomas during the night to explain why restraints were needed and to ask for his permission to use them on his mother. This may have prevented his reaction when he saw his mother sedated and physically restrained in the morning. Alternatively, a nurse or a member of the medical team could have explained to Thomas the reasons for the restraints as soon as he came in. By early morning the medical team would need to decide whether or not to renew the order for Kathleen's chemical and physical restraints, based on federal regulations. It may have been possible to remove the wrist straps at that time, since Kathleen seemed to be heavily sedated.

Thomas' statement that his mother should not be restrained without his express permission should not end the discussion. He may see himself as protecting his mother's dignity. His reaction may also stem from fear and anxiety about his mother's condition and uncertainty about her ability to recover. Understanding his emotional response, the medical team should have a conversation with Thomas to fully inform him about her diagnosis and prognosis, and the possibility that restraints may be needed in the future, if she is to be treated appropriately. This will be particularly important if Thomas needs to make medical decisions for his mother at some later time.

If Kathleen is considered capable of making her own medical decisions once she recovers from the effects of the sedation, she should be informed about the events of the previous night and the reasons for the use of restraints. Based on the *Principle of Veracity*, she needs to have complete information about her medical condition, the expected outcome, and the various medications she is receiving. Since her delirium might return, she should be asked to give consent to sedation and physical restraint if these become necessary in the future. Her agreement should be noted in her medical chart. Based on the *Principle of Respect for Autonomy*, her consent could then be used to authorize the use of restraints in the future, even if Thomas disagrees.

The Formulation

Now that the evidence-based medicine, legal precedent, and relevant ethical principles for this case have been reviewed, formulate a strategy to address the ethical conflicts in this case. If necessary, perform additional research into local and state laws and hospital regulations. Consider delving further into the background medical literature to assist with making sound therapeutic decisions. Devise a treatment approach that addresses the needs of the patient and her family, that is both ethically and medically sound, and that is culturally competent. Ensure that the strategy employs fair and appropriate utilization of medical resources, and that the approach is practical and feasible within the limits of the medical system at large. Work out a clear and professional way to communicate the proposal to the patient and her family. Attempt to foresee challenges that may arise in conveying or implementing the plan. Determine what follow-up will be necessary to ensure that the chosen strategy remains successful for the patient in the long-term. Reflect on how the knowledge and skills learned from this case can be used to improve the care of patients that may be encountered in future practice.

Afterthoughts

In this case, chemical and physical restraints were used to help prevent a patient from self-harm when she was not in a coherent or competent state. Practitioners must use restraints in a responsible way and only when other, less restrictive methods of maintaining patient safety have been duly considered and deemed inappropriate, or if these methods have been implemented and failed. For example, using bedrails and verbal redirection to prevent a confused patient from falling out of their hospital bed is preferable to using wrist restraints. Additionally, the use of any form of restraints should be discontinued as soon as they are no longer necessary to maintain patient safety.

Would physical or chemical restraints be ethically justifiable if an oriented patient were having a violent outburst in the hospital? What circumstances make the use of restraints on patients ethical or not ethical?

Annotated References/Further Information

Code of Medical Ethics of the American Medical Association: Current Opinions with Annotations, 2006–2007 Edition. Council on Ethical and Judicial Affairs. Annotations prepared by the Southern Illinois University Schools of Medicine and Law. The subsection on Use of Restraints (8.17) includes a helpful list of guidelines for medical practitioners.
Medicare and Medicaid Programs; Hospital Conditions of Participation: Patients' Rights; Final Rule (2006). 42 Code of Federal Regulations Part 482. 482.13 (e) Standard: Restraint or seclusion.

Snyder L, JD and Leffler C, JD. Ethics Manual, Fifth Edition. Ethics and Human Rights Committee, American College of Physicians. Annals of Internal Medicine. 142(7):560–582, 5 April 2005.

http://www.cms.hhs.gov/CFCsAndCoPs/downloads/finalpatientrightsrule.pdf. Accessed on October 15, 2007.

Fick DM et al. Delirium superimposed on dementia: a systematic review. Journal of the American Geriatrics Society. 50(10):1723–32, 2002 Oct.

Young J and Inouye SK. Delirium in older people. BMJ. 334(7598):842–6, 2007 Apr 21.

Case 7
When Patient Behavior Constitutes Abuse or Neglect

The Patient

Angela G. is a 33-year-old female with a history of tobacco dependence and asthma, who presented to the Emergency Department with a three day history of dyspnea, wheezing, and productive cough. She additionally noted subjective fevers and chills, and reported having sick contacts at home with upper respiratory infections. Her dyspnea and wheezing did not improve with use of her bronchodilator inhaler at home, prompting her to come to the hospital. The patient reported a 17 pack-year history of cigarette smoking, and denied alcohol or illicit drug use. There was no relevant family history. She is a single mother to two daughters, aged five- and seven-years old.

Her physical exam was notable for mild respiratory distress, a low-grade fever, and diffuse bilateral wheezes. A hematological and basic metabolic laboratory work-up was unrevealing. Chest radiography and electrocardiography were within normal limits. A urine toxicology screen was positive for cocaine and opiates. She was admitted for treatment of her presumed asthma exacerbation.

The Ethical Dilemma

Based on the urine toxicology screen, the admitting medical team asked Angela again about illicit drug use. She said she had used drugs only "recreationally" in the past, and denied ever using intravenous drugs. However, she said her drug use had somewhat increased recently since her new boyfriend occasionally sold drugs to "make a living" and, therefore, the drugs "were around more." After the team left Angela's room, one of the medical residents expressed her concern to the team about the presence of illegal drug activity in a home with young children present.

Questions for thought and discussion: Does the presence of illegal drugs in Angela's home constitute a situation of child abuse? What about the presence of illegal drug dealing? How are child abuse and neglect defined? Is the drug use or dealing in this case "reportable" to authorities?

From: *Evidence-Based Medical Ethics*
By: J.E. Snyder and C.C. Gauthier © Humana Press, Totowa, NJ

Question for thought and discussion: Is the team obligated to share their concerns about abuse/neglect with Angela herself, or do they make an anonymous report to child protective services?

Question for thought and discussion: Should Angela's medical team report the known illegal drug activity to the local authorities, regardless of the children's presence in the home?

Question for thought and discussion: Is performing a urine toxicology screen on Angela without her consent ethical, given the potential implications of a positive test?

The Medicine

Urine toxicology screens are not infallible. Several common medications, such as NSAIDs, may cause false-positive identification of illicit drugs on urine toxicology screens (see Table 7.1). However, such results are relatively uncommon, and cocaine is not one of the substances that is normally misidentified. Passively inhaled crack cocaine does not generally cause detectable levels in the urine either. Positive cocaine results on a urine toxicology screen generally suggest the last use was within the preceding 72 hours. It is important to note that a positive urine toxicology screen for cocaine does not differentiate between rare or one-time usage and chronic, addictive use. Also, if a patient adds some adulterant substances to urine this may also lead to false-negative urine toxicology screens (Table 7.2).

A number of studies have looked at the use of cocaine among patients with asthma. In 1996 Levenson, et al. determined that almost one-third of 92 asthma deaths in the Chicago area over an 18-month study period were confounded by alcohol or cocaine

Table 7.1 Some examples of medications causing false-positive urine toxicology screen results

Toxicology screen false positive for:	Medications/substances implicated in false result:
Amphetamine, methamphetamine	amantadine, bupropion, desipramine, diet-pills (usually of non-U.S. origin), ephedrine, fenfluramine, ma huang (a Chinese herb), methylphenidate, phentermine, phenylephrine, phenylpropanolamine, propylhexedrine, pseudoephedrine, ranitidine, selegiline, trazadone, Vicks inhaler™ (desoxyephedrine)
Barbiturates	NSAIDs (e.g. ibuprofen, naproxen)
Benzodiazepines	"black pearls" (tung sheuh, a Chinese herb), oxaprozin
Cannabinoids	dronabinol, efavirenz, hemp-containing foods, NSAIDs (e.g. ibuprofen, naproxen)
Opiates	chlorpromazine, codeine, fluoroquinolones, foods containing poppy seeds, quinine, rifampin
Phencyclidine	chlorpromazine, dextromethorphan, diphenhydramine, doxylamine, meperidine, thioridazine, venlafaxine

Table 7.2 Some examples of adulterants that, when added to urine, may cause false-negative urine toxicology (enzyme immunoassay) screen results

golden-seal tea
household bleach
household drain cleaner (e.g. Liquid Drāno™)
liquid hand soap
sodium chloride
vinegar
Visine™ eye drops

Source adapted from: Mikkelsen SL and Ash KO. Adulterants Causing False Negatives in Illicit Drug Testing. Clinical Chemistry. 34(11):2333–2336. 1 November 1988.

use. A later prospective study of adults with asthma exacerbation by Rome, et al. found a prevalence rate for cocaine use of 13 percent, and the cocaine users had a statistically significant increase in hospital admissions, compared to non-users. However, active tobacco smoking was additionally more prevalent in the cocaine users group. Additional studies suggest that increases in cocaine use in the United States, and specifically among patients with asthma, may be associated with increases in asthma morbidity.

The 2001 Substance Abuse and Mental Health Services Administration (SAMHSA) National Household Survey on Drug Abuse estimated that there are over 6 million children living with one or more parents that abuse or are dependent on alcohol or illicit drugs. A number of deleterious effects have been noted on children whose parents abuse illicit drugs. Although not definitively proven to be teratogenic or affecting early development, cocaine use during pregnancy is associated with higher rates of spontaneous abortion and abruption. Children of illicit drug-using parents are more likely to experience neglect, physical abuse, and sexual abuse in their lives. They tend to have poorer academic performance and social adjustment skills, and may struggle later in life when they become parents themselves due to the lack of strong role models in their past. They are also more likely to become substance abusers themselves. Exposure to situations of poverty, criminal behavior, parental absence, and parental imprisonment may all contribute to these observed phenomena.

The Law

Although reporting abuse or neglect is usually done at the local, county, or state level, it is federal and state law that defines these entities. The Child Abuse Prevention and Treatment Act (CAPTA) is federal legislation that defines child abuse or neglect as "any recent act or failure to act on the part of a parent or care-taker, which results in death, serious physical or emotional harm, sexual abuse, or exploitation, or an act or failure to act which presents an imminent risk of serious harm" (42 U.S.C.A. § 5106g(2) (West Supp. 1998)). Substance abuse by a child's caretakers is considered to be potentially harmful to the child's mental and physical

Table 7.3 United States laws relating illegal drugs to child abuse

States where it is a felony to manufacture or possess methamphetamines in the presence of a child
Georgia
Illinois
Nebraska
New Hampshire
Pennsylvania
Virginia
West Virginia
Wyoming
States where it is a felony to manufacture or possess any controlled substance in the presence of a child
Idaho
Louisiana
Ohio
States where additional penalties enhance convictions for the manufacture of methamphetamines with a child on the premises
California
Mississippi
Montana
North Carolina
Washington (state)
States where it is considered child endangerment to expose children to the manufacture, possession, or distribution of illegal drugs
Alaska
Iowa
Kansas
Minnesota
Missouri
States where it is a crime to expose a child to illegal drugs or drug-related paraphernalia
North Dakota
Utah
States where it is a felony to sell or give illegal drugs to a child
North Carolina
Wyoming

Source adapted from: *http://www.childwelfare.gov/systemwide/laws_policies/statutes/define.cfm.* Accessed October 15, 2007.

health, and can additionally threaten their safety. Whereas 45 states and the District of Columbia have laws within their child protection statutes that specifically address parental abuse of illegal drugs, only 23 states have laws defining substance abuse in terms of child abuse or neglect (see Table 7.3). Fifteen states and the District of Columbia have reporting procedures in place if there is proof that a newborn has been prenatally exposed to drugs, alcohol, or other controlled substances. Thirteen states and the District of Columbia include prenatal exposure to drugs, alcohol, or other controlled substances in their definitions of child abuse or neglect.

Anyone who suspects that a child is being abused or neglected is encouraged to notify their local law enforcement or child protection authorities, and all states have some legal statutes that mandate this reporting by certain individuals. For example, most states identify persons of certain professions with frequent contact with children, including those in the medical and mental health fields, as mandatory reporters of abuse. Penalties can be associated with non-reporting of abuse or neglect in these states. Confidential protection of the reporting person's identity is maintained in 39 states and the District of Columbia, and this information is generally discoverable only with the reporter's consent or in other, very specific circumstances. However, these rules can vary from state to state.

Although law enforcement agencies encourage individuals to report any known illegal activity, there are no specific laws regarding mandatory reporting of illegal drug activity that is not placing children or other vulnerable persons at risk. Currently laws about illegal drugs generally focus on their manufacture, importing and exporting, distribution, and possession. When a patient admits illegal drug use to a medical or mental health professional, it is considered privileged information, and patients must feel that they can be forthright with their provider and know that such information will be kept confidential. Without complete information, the provider may be unable to render appropriate care to the patient. Physician-patient privilege becomes more legally complicated when a patient threatens bodily harm to an individual. In these cases a medical or mental health professional is obligated to take appropriate steps to protect the intended victim, by reporting the plan to harm to the intended victim or to local law enforcement. The legal precedent for this was established by the landmark case of *Tarasoff v. Regents of the University of California*, where a patient of a psychologist at the University of California – Berkeley committed murder after telling his psychologist that he planned to do so. The Supreme Court of California determined that there was a duty to protect the intended victim that superseded the duty to maintain the patient's right to confidentiality.

In the case of Angela's positive urine toxicology screen, if Angela admitted using illegal drugs in a home without children to her practitioner, then this information would be considered privileged and there is no law obligating the medical team to report the drug use to law enforcement. Information about the manufacture and distribution of illegal drugs in a home without children may be considered a murkier situation as one can argue that "bodily harm" may come to those who are given or sold the illegal drugs. Nevertheless, there is not any overt and stated threat by the drug dealer to injure a specific individual. However, if illegal drugs are manufactured or distributed in a home where there are small children, this does constitute a situation of child abuse/neglect by most legal standards and the medical team should report this information to the local child protective services agency. The agency will then conduct an investigation into the home situation and determine if the children are at risk and need to be removed from the caretaker's custody.

With regard to the legality of random toxicology screening of patients, there is little clear precedent dictating whether or not this is allowable for the purpose of providing safe and appropriate medical care. Hence, the general practice of ordering urine drug screens for medical purposes is considered legally acceptable. Drug screens are frequently performed on patients in the hospital setting, and often without the patient's consent. It is current opinion that general consent for medical care implies the consent for basic laboratory testing, including the use of toxicology screens. However, many experts feel that if the purpose of the toxicology screen is to involve child protective services in the case or to prosecute the patient, then the patient should provide specific informed consent for the test. The 2001 United States Supreme Court case of *Ferguson v. City of Charleston* illustrated a situation where the Medical University of South Carolina had entered an agreement with Charleston law enforcement officials whereby pregnant women who had positive urine screens for cocaine were subsequently arrested on charges of drug possession, distribution to a minor, or child neglect. Thirty women were arrested under this program over a five-year period. The Court ruled that, in this case, screening pregnant women for cocaine use without consent, warrants, or probable cause constituted a violation of the Fourth Amendment which protects individuals from unwarranted searches and seizures. Although the stated intention of the program, to improve prenatal and postnatal care of substance using expectant mothers, may have been a noble one, opponents to the concept of the policy state that it may actually work to discourage women from seeking perinatal care or treatment of substance use disorders.

In 2002, Abel and Kruger surveyed 847 obstetricians, pediatricians, and family practice physicians about their thoughts on mandatory reporting of pregnant women with active substance abuse to the police or child protection agencies. Of those surveyed, 61 percent felt that fear of prosecution would deter pregnant women with substance abuse problems from seeking prenatal care. However, a majority also believed that pregnant women have a moral and legal duty to ensure their babies were healthy, and that screening for HIV and substance abuse should be mandatory. More than half felt that laws about child abuse and neglect should be modified to include substance abuse during pregnancy, and that this should be considered as grounds for removing the child at birth from maternal custody. Only a minority of those surveyed (less than 31%) felt that the women should face criminal prosecution; however, a vast majority supported obligatory participation in a substance treatment program.

The Ethics

An ethical evaluation of urine toxicology screening, performed without patient consent, raises several questions. Is this test ordered for all patients who come to the Emergency Department or are admitted to the hospital? If not, when is it usually ordered? Why was it ordered for this particular patient? Was there reason to suspect drug abuse in this case?

If all patients who come to the Emergency Department or are admitted to the hospital undergo a toxicology screen, an argument can be made that it is ethical, based on the *Principles of Beneficence* and *Non-Maleficence*. A hospital policy mandating this screening test would allow practitioners to better diagnose their patients' medical conditions and make prescribing medications safer. Practitioners will be better able to promote the health of their patients if they are aware of recent drug use. Because some illegal drugs interact with prescribed medications, physicians could unintentionally cause harm to their patients by prescribing certain medications - a harm that could be avoided with prior knowledge about patients' drug use.

However, such a policy would appear to violate the *Principle of Respect for Autonomy* because patients are not giving voluntary informed consent prior to the screening test. Voluntary informed consent is necessary if patients are going to be permitted to make decisions about their own health care. Full information about recommended medical tests and treatments must be given to capable patients, they must understand this information, and they must agree to these medical interventions without coercion or undue pressure. Even if a hospital posted a policy of mandatory drug screening for all patients in the Emergency Department, with detailed information about the test in easily understandable language, patients in dire need of emergency care would literally have no choice but to submit to the screening test. This would violate the requirement of voluntary agreement.

If the hospital does not screen all patients for illicit drugs, it will be important to justify screening only particular patients. Perhaps the medical team suspected drug abuse and when Angela denied it and they felt justified using the drug screen to settle the issue. They may be able to use the *Principles of Beneficence* and *Non-Maleficence* to argue that they needed this information to provide the best medical care to Angela, as noted above. However, performing this test on Angela, without her knowledge or consent, violates the *Principle of Respect for Autonomy*.

It might be argued that patients give implied consent for any procedures emergency physicians believe are in the patient's best interests, simply by seeking medical care in the Emergency Department. The concept of "implied consent," however, is ethically suspect when it is applied to screening tests that may have negative consequences for the patient. This would be true of HIV screening as well as screening for illicit drugs. In both cases, although the results of these screening tests would help practitioners provide better medical care to their patients, this information could also harm their patients in other ways.

By requiring voluntary informed consent for medical tests and treatments, the patient can compare the risks of harm to the potential benefit of these procedures. Performing HIV and drug screening tests without fully informed, express consent, denies patients the opportunity to compare the value of this information for the medical care that can be provided to them with the risks of social harm the information may pose.

However, patients who use illicit drugs may not seek emergency medical care when needed if drug screening is being done without their knowledge and consent when they come to the Emergency Department. Such a policy could exclude a particularly vulnerable segment of the population from receiving emergency medical treatment.

There may be a way to provide the medical benefits of drug screening, without violating the *Principle of Respect for Autonomy*. Hospital policy could allow emergency physicians to order a toxicology screen on a routine basis whenever they suspect illegal drug use. When this occurs, the physician could inform the patient of this policy, briefly explain the medical benefits of the test results and give the patient the chance to "opt out" of screening. Most importantly, it must be made clear that opting out will not deny a patient emergency medical treatment. This is necessary to avoid pressuring people into submitting to the drug screening to get emergency care.

If a patient opts out of toxicology screening, that means the patient has chosen to forgo the benefits that the results of the test would have for that patient's own health care. It will be important for practitioners to remember that the benefits of the screening are for the individual patient, not for the society, since individual drug use is not a reportable event. This may make it easier for them to accept a patient's decision to opt out of routine drug screening.

The Formulation

Now that the evidence-based medicine, legal precedent, and relevant ethical principles for this case have been reviewed, formulate a strategy to address the ethical conflicts in this case. If necessary, perform additional research into local and state laws and hospital regulations. Consider delving further into the background medical literature to assist with making sound therapeutic decisions. Devise a treatment approach that addresses the needs of the patient and her family, that is both ethically and medically sound, and that is culturally competent. Ensure that the strategy employs fair and appropriate utilization of medical resources, and that the approach is practical and feasible within the limits of the medical system at large. Work out a clear and professional way to communicate the proposal to the patient and her family. Attempt to foresee challenges that may arise in conveying or implementing the plan. Determine what follow-up will be necessary to ensure that the chosen strategy remains successful for the patient in the long-term. Reflect on how the knowledge and skills learned from this case can be used to improve the care of patients that may be encountered in future practice.

Afterthoughts

In this case a positive urine screen for illicit drugs in a patient presenting to the emergency department with an asthma exacerbation could potentially result in her losing custody of her children. For this reason, one may ethically question if patients should be "screened" for drug use at all, if consent is not obtained first. The use of certain illicit substances is important for a health care provider to know since patients such as Angela do not always disclose drug use to their provider, and knowing about the drug use can be essential to the patient's care. Illicit drug use puts patients at higher risk for infections such as HIV and hepatitis C, and screening tests can help quickly identify patients at risk for these and other diseases. Additionally, practitioners need to know about the interactions of prescribed medications and illicit drugs. For example, administering medications such as beta-blockers can be potentially dangerous to patients with cocaine in their bloodstream. If a patient does not admit to illicit drug use, and does not consent to a urine drug screen, it can limit the practitioner's ability to appropriately treat that patient.

What is it about certain diagnostic tests that make it necessary to obtain patient consent before ordering them? It is impractical to explain and obtain consent for every diagnostic study ordered on a patient (e.g., a serum chloride level), but ethically necessary to obtain consent for certain ones (e.g., an HIV screen) due to the implications of the results. What level of *implied* consent is there when a patient comes to an emergency room for help with a medical problem?

Annotated References/Further Information

Code of Medical Ethics of the American Medical Association: Current Opinions with Annotations, 2006–2007 Edition. Council on Ethical and Judicial Affairs. Annotations prepared by the Southern Illinois University Schools of Medicine and Law.

Snyder L, JD, and Leffler C, JD. Ethics Manual, Fifth Edition. Ethics and Human Rights Committee, American College of Physicians. Annals of Internal Medicine. 142(7):560–582, 5 April 2005.

Vincent EC et al. What common substances can cause false positives on urine screens for drugs of abuse? The Journal of Family Practice. 55(10):893–897.

Verstraete AG. Detection times of drugs of abuse in blood, urine, and oral fluid. Therapeutic Drug Monitoring. 26(2):200–205. April 2004.

Osterloh JD and Becker CE. Chemical dependency and drug testing in the workplace. The Western Journal of Medicine. 152(5):506–513. May 1990.

Katz N and Fanciullo GJ. Role of rrine toxicology testing in the management of chronic opioid therapy. The Clinical Journal of Pain. 18(4, Supplement):S76–S82. July/August 2002.

Narcessian EJ and Yoon HJ. False-Positive urine drug screen: beware the poppy seed bagel. Journal of Pain and Symptom Management. 14(5):261–263. November 1997.

Kipinis S et al. Alcohol & drug screens. A guide to the interpretation and effective use of screens for substances of abuse. NYS Office of Alcoholism and Substance Abuse Services – Addiction Medicine Series. Updated 4/2007. Acquired from: *http://www.oasas.state.ny.us/AdMed/documents/drugscreen.pdf*. Accessed October 15, 2007.

Mikkelsen SL and Ash KO. Adulterants causing false negatives in illicit drug testing. Clinical Chemistry. 34(11):2333–2336. 1 November 1988.

Foley EM. Drug screening and criminal prosecution of pregnant women. Journal of Obstetric, Gynecologic, and Neonatal Nursing. 31(2):133–137. March/April 2002.

Rome LA et al. Prevalence of cocaine use and its impact on asthma exacerbation in an urban population. Chest. 117(5):1324–1329. May 2000.

Levenson T et al. Asthma deaths confounded by substance abuse. Chest. 110:604–610. 1996.

The NHSDA Report: Children living with substance abusing or substance dependent parents, from SAMHSA's National Household Survey on Drug Abuse. Acquired from: *http://www.drugabusestatistics.samhsa.gov/2k3/children/children.cfm*. Accessed October 15, 2007.

Keen J and Alison LH. Drug misusing parents: key points for health professionals. Archives of Disease in Childhood. 2001;85:296–299.

Kumpfer KL. Outcome measures of interventions in the study of children of substance-abusing parents. Pediatrics. 103(5 Pt 2):1128–44, 1999 May.

http://www.childwelfare.gov/systemwide/laws_policies/statutes/define.cfm. Accessed October 15, 2007.

National Drug Policy: United States of America. Prepared For The Senate Special Committee On Illegal Drugs. Dolin B, Law and Government Division. 24 July 2001, Library of Parliament. *http://www.parl.gc.ca/37/1/parlbus/commbus/senate/Com-e/ille-e/library-e/dolin2-e.htm.* Accessed October 15, 2007.

http://www.stanford.edu/group/psylawseminar/Tarsoff%20I.htm. Accessed October 15, 2007.

Beauchamp TL and Childress JF. Principles of Biomedical Ethics, Fifth Edition, Oxford University Press, 2001. On pages 65–67 there is an interesting discussion of different kinds of consent, including implied consent.

http://www.reproductiverights.org/crt_preg_ferguson.html. Accessed October 15, 2007.

http://www.doh.wa.gov/cfh/mch/documents/screening_guidelines.pdf. Accessed October 15, 2007.

http://www.doh.wa.gov/cfh/mch/documents/Hospital_guidelines_drugexposed_newborns.pdf. Accessed October 15, 2007.

Abel EL and Kruger M. Physician attitudes concerning legal coercion of pregnant alcohol and drug abusers. American Journal of Obstetrics and Gynecology. 186(4):768–772. April 2002.

Case 8
When a Patient Has "Burned His Bridges"

The Patient

Raymond D. is a 36-year-old male who became paraplegic eight years prior to admission when, while driving under the influence of alcohol and cocaine, he lost control of the vehicle he was driving, crashed into a tree and sustained a lower spinal cord injury. He was left wheelchair-bound by the accident and must perform twice-daily self-catheterization for his neurogenic bladder and resulting urinary retention. In the subsequent years Raymond continued to abuse cocaine and was additionally diagnosed with major depression and antisocial personality trait. He had multiple hospital admissions for urosepsis, and developed a non-healing decubitus ulcer in the sacral area.

Raymond's home situation was tumultuous as well, and he gradually was disowned by all his family members. Following his frequent hospitalizations, he often had no place to be discharged to. Due to his paraplegia and wound care needs he was usually discharged to a skilled nursing facility, but then later would sign out of the facility and temporarily move in with an acquaintance or into an apartment.

At the time of his current admission, Raymond had been living with a friend after being recently evicted from an apartment complex. He presented to the Emergency Department after calling Emergency Medical Services with complaints of fevers, chills, diaphoresis, generalized weakness, and foul-smelling urine. His home medications included muscle relaxants for treating his chronic spasticity, and some narcotic analgesic agents for chronic back and lower extremity pain. Raymond admitted to smoking one pack of cigarettes per day, but denied active alcohol or illicit drug use. On physical exam, he was febrile and had borderline hypotension. His sacral decubitus ulcer had clear signs of wound infection. Laboratory evaluation revealed an elevated white blood cell count and a urinary tract infection. A urine toxicology screen was positive for cocaine. Blood and urine cultures were sent, broad-spectrum antibiotics started, and he was admitted to the hospital for treatment of both wound and urinary tract infections.

From: *Evidence-Based Medical Ethics*
By: J.E. Snyder and C.C. Gauthier © Humana Press, Totowa, NJ

The Ethical Dilemma

Raymond's low blood pressure responded quickly and favorably to IV fluid administration and antibiotics. Plastic surgery was called to debride his ulcer, and appropriate wound care was performed. He had become slightly deconditioned with regard to his upper extremity strength during the hospitalization, and physical therapy worked with him on a daily basis. When he was ready to be discharged from the hospital, Raymond's friend refused to let him stay at his apartment anymore. Several applications were placed by the case management team to skilled nursing facilities in the area for post-hospital care. Every application was rejected, with facilities unanimously citing past negative interactions with the patient due to behavioral problems and ongoing substance abuse. In fact, every facility in a three-county area refused to accept the patient. Similarly, all the local home health care agencies refused to assist with his care, making discharge to an apartment or even motel room impossible as well. Despite being medically stable for discharge, the patient had no reasonable plans for disposition.

Question for thought and discussion: Is it the hospital's responsibility to ensure a safe discharge plan for a patient that is *persona non grata* at every potential site of disposition?

Raymond remained in the hospital for another two weeks while the case management team feverishly worked to find a suitable disposition for him. Eventually, a skilled nursing facility 120 miles away from the hospital agreed to accept him on discharge. However, Raymond refused to be discharged so far away from his hometown.

Question for thought and discussion: Can Raymond refuse to accept the only available option for a safe discharge that the hospital can provide? Is it the hospital's responsibility to keep caring for him on site if he refuses discharge?

The Medicine

The annual incidence of spinal cord injuries (SCI) in the United States is not definitively known, but excluding those cases where the individual dies at the scene of the inciting accident, it is estimated that there are 11,000 new cases annually. There are about one-quarter million Americans currently alive with history of SCI. Most commonly, SCI injuries occur in men (77.8% of cases), at an average age of 38, and motor vehicle accidents account for almost half of the injuries. SCI can occur at any level of the spinal cord, with incomplete tetraplegia and complete paraplegia being the most common results (34.1% and 23.0% of cases, respectively), followed by incomplete paraplegia (18.5%) and complete tetraplegia (18.3%).

The vast majority of patients with SCI who survive their injuries are discharged to their home after care for their initial injury, and not an institution. Ten years after the inciting accident, 32.4 percent of individuals with paraplegia and 24.2 percent of those with tetraplegia are employed. The lifetime health care and living-related costs of sustaining a SCI at age 25 are almost $1 million for individuals with paraplegia and almost $3 million for those with "high" (C1-C4) tetraplegia. Many secondary impairments may affect individuals with SCI, including problems such as urinary tract infections, spasticity, hypotension, autonomic dysreflexia, pressure ulcers, overuse syndromes, and venous thrombosis. In one study by Proulx, et al. urinary tract infection was the most prevalent of these impairments, with 56 percent of patients experiencing UTI over the preceding 12 months. The prevalence of pressure ulcers in this study was 28 percent.

Life expectancy is lowered after sustaining SCI, often due to developing complications such as pneumonia, pulmonary embolism, or septicemia. However, this is an area of improvement in recent years. As one example, and based on estimates from the National Spinal Cord Injury Statistical Center (NSCISC), a 40-year-old who survives one year after the inciting injury may expect to survive, on average, an additional 28.6 years with paraplegia or 21.6 years with high tetraplegia.

Data do not support a correlation between physical disabilities such as paraplegia or tetraplegia and the development of personality disorders. One study by Bockian, et al. determined that the incidence of personality disorders in a population of Veterans Administration (VA) patients with SCI was no greater than matched controls without SCI. However, this study and several others have in fact noted a high incidence of depression among individuals with SCI. Some studies of substance abuse among individuals with SCI indicate higher rates of alcohol and marijuana usage, as well as more frequent misuse of prescription drugs. The higher rate of substance abuse in these studies is often associated with higher rates of depressive symptoms. The chronic pain often experienced by individuals with SCI, estimated to occur with a prevalence as high as 47 to 96 percent, may also contribute to this problem.

The Law

Once a patient like Raymond is medically cleared for discharge by his medical team, Medicare and most other insurance carriers will not reimburse for costs related to further hospitalization. Therefore, it is in a hospital's financial best interest to efficiently find a patient a suitable disposition when that patient has been declared medically stable. Patients that are medically stable and without special nursing, rehabilitation, or other skilled needs can be discharged from the hospital at that time, even if they do not agree with the discharge plan. The hospital does not need to provide further assistance to them. For patients that do have special skilled needs at the time of discharge, a reasonable effort must be made to accommodate them and provide a safe disposition. If the patient refuses to accept the discharge plan that the hospital offers them, then the hospital has fulfilled their legal obligations to the patient and may proceed with a discharge.

If a patient is mentally or physically disabled, such as Raymond, reasonable accommodations need to be made to ensure a safe disposition. If a disabled patient feels that health or human services have been denied to them as a result of their disability, they can formally file a complaint with their regional Office of Civil Rights, a branch of the U.S. Department of Health and Human Services. However, similar to a patient without a disability, if a reasonable effort has been made to accomplish a safe disposition and the patient refuses to accept the plan, the hospital has probably fulfilled their legal obligations. Issues, such as the ones surrounding Raymond's discharge plan, can be complicated however, and the hospital's risk management and legal counsel should likely be consulted prior to enforcing a discharge plan on any patient who does not accept it.

The Ethics

The case management team's efforts to find a place for Raymond to go after discharge are admirable. The motivation for these efforts can be found in the *Principles of Beneficence* and *Non-Maleficence*. The hospital felt responsible for providing a safe place for Raymond to receive post-hospital care so that his continued recovery would be successful. They also wanted to prevent the harm that could have come to him if he had not had an appropriate place to go after discharge. The hospital was trying to avoid causing harm to Raymond by discharging him into the streets, knowing his condition could worsen and he could die without proper care.

The hospital has fulfilled its ethical responsibility to Raymond in two ways. First, he was allowed to remain in the hospital for two weeks, after he was medically stable, while a safe place for discharge was located. In addition, the hospital case management team worked tirelessly to find a skilled nursing facility willing to accept him for post-hospital care.

Raymond refuses to be discharged to this facility because it is too far away. The *Principle of Respect for Autonomy* allows patients to accept or refuse recommended medical procedures. However, it does not mean that patients can demand medical treatment, e.g., continued hospital care that they no longer need. According to this principle, Raymond can accept or refuse the hospital's recommendation for placement after discharge. While Raymond can refuse the placement, he needs to understand that he can no longer stay at the hospital and that if he refuses this placement, he needs to find a place to go following discharge. It is not the hospital's responsibility to continue to look for a placement solution that is more acceptable to Raymond.

The hospital also does not have an ethical responsibility to keep caring for Raymond on site now that he is medically stable and an appropriate facility for discharge has been located and is willing to accept him. The *Principle of Distributive Justice* concerns the fair distribution of health care resources. It could be argued that Raymond's continued stay in the hospital is not a fair use of health care resources and may be depriving other patients of needed hospital care. If the ideal basis for distribution is need, once Raymond no longer needs the level of care provided at the hospital, he should no longer have access to those resources.

The Formulation

Now that the evidence-based medicine, legal precedent, and relevant ethical principles for this case have been reviewed, formulate a strategy to address the ethical conflicts in this case. If necessary, perform additional research into local and state laws and hospital regulations. Consider delving further into the background medical literature to assist with making sound therapeutic decisions. Devise a treatment approach that addresses the needs of the patient and his family, that is both ethically and medically sound, and that is culturally competent. Ensure that the strategy employs fair and appropriate utilization of medical resources, and that the approach is practical and feasible within the limits of the medical system at large. Work out a clear and professional way to communicate the proposal to the patient and his family. Attempt to foresee challenges that may arise in conveying or implementing the plan. Determine what follow-up will be necessary to ensure that the chosen strategy remains successful for the patient in the long-term. Reflect on how the knowledge and skills learned from this case can be used to improve the care of patients that may be encountered in future practice.

Afterthoughts

In this case, a patient with a physical disability had "burned bridges" with facilities and companies providing out-of-hospital care, which limited the hospital's ability to provide a safe discharge plan for him. How would this case have been different if Raymond's disability had been a cognitive or psychiatric one, instead of physical? What if the patient was a disabled child, and his or her parents refused to accept the hospital's discharge plan?

Annotated References/Further Information

Code of Medical Ethics of the American Medical Association: Current Opinions with Annotations, 2006–2007 Edition. Council on Ethical and Judicial Affairs. Annotations prepared by the Southern Illinois University Schools of Medicine and Law.

Snyder L, JD, and Leffler C, JD. Ethics Manual, Fifth Edition. Ethics and Human Rights Committee, American College of Physicians. Annals of Internal Medicine. 142(7):560–582, 5 April 2005.

National Spinal Cord Injury Statistical Center (NSCISC). Facts and Figures at a Glance. June 2006. Acquired from: *http://www.spinalcord.uab.edu/show.asp?durki=21446*. Accessed October 15, 2007.

Bockian NR et al. Personality disorders and spinal cord injury: a pilot study. Journal of Clinical Psychology in Medical Settings. 10(4):307–313.

Tate DG et al. Patterns of alcohol and substance use and abuse in persons with spinal cord injury: risk factors and correlates. Archives of Physical Medicine and Rehabilitation. 85(11):1837–1847, November 2004.

Heinemann AW et al. Prescription medication misuse among persons with spinal cord injuries. International Journal of the Addictions. 27(3):301–316. March 1992.

SAMHSA's National Household Survey on Drug Abuse. Acquired from: *http://www.samhsa.gov/ news/newsreleases/060907_nsduh.aspx*. Accessed October 15, 2007.

Yezierski RP. Pain following spinal cord injury: the clinical problem and experimental studies. Pain. 68:185–94, 1996.

Noreau L et al. Secondary impairments after spinal cord injury. A population-based study. American Journal of Physical Medicine & Rehabilitation. 79(6):526–535, 2000.

http://www.hhs.gov. Accessed October 15, 2007.

Case 9
When a Patient Is Administratively Discharged

The Patient

Dennis P. is a 27-year-old male with a history of ulcerative colitis that was diagnosed at age 17 when he underwent colectomy for acute toxic megacolon and bowel perforation. Over the next several years he had recurrent episodes of partial small bowel obstruction that required hospitalization, and underwent three exploratory laparotomies for lysis of adhesions. Recently, Dennis was coming to the Emergency Department up to twice per month with complaints of abdominal pain. Each visit resulted in radiographic studies of his abdomen showing postoperative changes that were difficult to interpret for acute illness. He would be given a diagnosis of recurrent partial small bowel obstruction and admitted to the general surgery service for observation, placement of a nasogastric tube, bowel rest, intravenous fluids, and analgesia. His attitude was often belligerent towards hospital staff, and he loudly and frequently voiced demands and complaints. Two to three days into his hospital stay, Dennis' symptoms would usually spontaneously improve and he would be discharged home with a limited amount of pain medication. He developed a reputation among the surgical residents as a "drug seeker" for these repeated similar admissions that required minimal intervention.

At the time of the current presentation, Dennis noted a two-day history of worsening generalized abdominal pain, nausea, and loose bowel movements. He attributed his symptoms to "another bowel obstruction." His physical exam was notable for normal vital signs and a diffusely tender abdomen with guarding, but no rebound. Laboratory evaluation was unremarkable. CT scan of the abdomen was unchanged from multiple previous studies.

The Ethical Dilemma

Dennis' current presentation to the Emergency Department occurred on a particularly busy evening for the admitting surgical resident. The resident told the ED physician that Dennis' complaints were unchanged from any of his previous

From: *Evidence-Based Medical Ethics*
By: J.E. Snyder and C.C. Gauthier © Humana Press, Totowa, NJ

admissions and that "we don't have anything surgical to offer him." The resident refused to admit Dennis to her service, and stated "just have him follow up in surgical clinic."

Questions for thought and discussion: Can a physician refuse to admit a patient? Under what circumstances?

Subsequently, Dennis began making frequent "sick visit" appointments at the surgical clinic as well as Emergency Department visits. For concern of narcotic abuse, the surgical residents had him sign a contract that stated he would only receive a specified amount of narcotic medications per month and no further medication would ever be prescribed. The contract also stated that if he were found to be using any illegal street drugs or receiving prescription narcotics from other sources, he would no longer receive any prescription narcotic medications from the clinic. The contract also required Dennis to have a urine toxicology screen done whenever he requested narcotic medications.

Questions for thought and discussion: Are narcotic contracts ethical? If pain is a subjective complaint, how can one know that a patient may or may not need more pain medication than what is prescribed?

Question for thought and discussion: Is it ethical to require Dennis to give a urine sample for toxicology screening with each request for a prescription?

At one of Dennis' clinic visits where he requested refills for his pain medications, Dennis' toxicology screen came back positive for benzodiazepines and cocaine. The surgical resident seeing Dennis terminated his pain contract due to his illicit drug use. She also asked her supervising attending if they could "administratively discharge" Dennis from the clinic due to his rude attitude, narcotic-seeking behavior, and illicit drug use.

Questions for thought and discussion: Can a patient be discharged from a medical practice because a physician no longer wants to care for them?

The Medicine

Patients presenting to their health care provider with physical complaints that correlate to no specific organic pathology is common. These symptoms can become chronic and result in multiple visits to the provider, eliciting feelings of frustration from both parties. Patients may feel rejected by the provider, that not enough time is devoted to their care, or they may become frustrated by the lack of a distinct diagnosis. They may perceive that they are misunderstood, not taken seriously, or not believed. On the other hand, the provider may become frustrated with the demands of the patient, perceiving them as difficult or that they over-utilize medical resources.

Chronic non-cancer pain is a common problem in the United States, with an estimated prevalence among several studies of 20 to 60 percent of general medical patients. Living with chronic pain is difficult for patients as it impairs their ability to sleep, function within their family units, and may have implications for other disease states. Patients with chronic pain are also more likely to suffer from anxiety and depression. Chronic opioid therapy is often used to treat chronic non-cancer pain in an effort to improve a patient's quality of life. Challenges in using narcotics for long-term pain management include developing tolerance, decreased treatment efficacy, and potential for toxicity and abuse. Estimates of addiction risk for prescription opioids vary between 5 and 19 percent with different studies.

Abuse of prescription medications, particularly narcotic painkillers, is a rapidly growing problem in the United States. In 2002 there were 43,100 admissions for substance abuse treatment related to prescription narcotics. Although this number comprises only about 4 percent of all dependence-related admissions, it is more than double the number of cases from 10 years prior. Despite narcotic contracts and random urine toxicology screens, prescription narcotics are frequently misused. One study of 186 patients in a treatment center in West Virginia identified 355 narcotic contract violations over a period of one year, even though there was a well-defined protocol for narcotic prescription distribution and careful screening methods to protect against abuse. It is, nonetheless, appropriate to establish such guidelines for managing patients with chronic pain. Although false-positive and false-negative results are possible with urine toxicology screening (please refer back to Case 7, Tables 7.1 and 7.2), it may be a practical tool to help the practitioner assess inappropriate drug use in their patients.

The Law

The surgical resident's refusal to admit Dennis to the hospital may be legal under the federal Public Health Service Act. In its section on emergency services the act states, "A facility may discharge a person that has received emergency services or may transfer the person to another facility able to provide necessary services, when the appropriate medical personnel determine that discharge or transfer will not subject the person to a substantial risk of deterioration in medical condition" (42 Code of Federal Regulations Part 124.603 (b)). In this case emergency service provided through a physical exam, lab tests, and a CT scan determined that Dennis was not in immediate danger and did not require surgical services.

However, the question may be raised about Dennis' need for other services. It may have been better for an attending physician to review Dennis' results and decide if he should be admitted to the hospital. This would avoid any potential controversy over who was the "appropriate medical personnel" in this situation.

Physicians can discharge a patient from a medical practice, as long as the discharge does not constitute "patient abandonment." To avoid the charge of abandonment, physicians should take steps to ensure continuity of care for the patient. These include giving notice, preferably in writing, to the patient or the patient's family with enough time to permit another practitioner to take over the patient's care. The physician should also offer to transfer the patient's medical records to another practitioner of the patient or family's choice.

Physicians should consult hospital policies on admitting practices and discharging a patient from a hospital clinic. Individual state laws on discharge from the Emergency Department and on patient abandonment should also be consulted since these may vary from state to state.

The Ethics

Narcotics contracts could be justified using the *Principle of Non-Maleficence*. In this case the surgical residents need to limit the amount of narcotic medication that Dennis can receive each month, and they need to know that he is not using illegal drugs or obtaining drugs elsewhere to avoid harmful, dangerous drug interactions or overdose. Requiring a toxicology screening test could be justified in the same way. By confirming that Dennis is honoring his contract by testing for drugs in his system, the residents may feel more assured that their narcotics prescriptions will not cause a life-threatening drug interaction or overdose.

The narcotics contract and the required toxicology screen can also be justified based on the *Principle of Respect for Autonomy*. This principle does not permit patients to make demands of their physicians for medical treatments, including medications. Patients must be permitted, however, to accept or refuse the recommendations their physicians make. The residents have offered Dennis the option of receiving limited narcotic medications under certain conditions designed to protect him from a harmful drug overdose or interaction. Dennis is free to accept or refuse this offer.

An argument could be made that Dennis' consent to the contract is not voluntary due to the level of his pain. This may be a persuasive argument because of the subjectivity of pain. Although the residents believe that Dennis is asking for more narcotics because he is a "drug seeker," he could be experiencing more pain than others with his condition might experience. Because this is impossible to objectively determine, the best solution is for the residents to monitor Dennis' condition closely, provide him with a reasonable amount of narcotics, and see how he does with the contract. If the residents later feel that the prescribed narcotic dosages are not adequate, they can be adjusted as needed.

A patient may be discharged from a medical practice as long as arrangements are made for the continuity of the patient's care. In this case, the surgical resident was correct to ask her supervising attending to consider this, rather than doing it herself. She might also have consulted with her fellow residents before making the decision to terminate Dennis' pain contract.

One other avenue that the residents might consider is to ask a hospital social worker to meet with them and then Dennis to discuss drug rehabilitation services in the community. While the residents seem to be focused on Dennis' present drug seeking behavior and his rude and belligerent attitude, they need to remember that his narcotic dependence is the result of his originally being treated for a legitimate medical condition.

The Formulation

Now that the evidence-based medicine, legal precedent, and relevant ethical principles for this case have been reviewed, formulate a strategy to address the ethical conflicts in this case. If necessary, perform additional research into local and state laws and hospital regulations. Consider delving further into the background medical literature to assist with making sound therapeutic decisions. Devise a treatment approach that addresses the needs of the patient and his family, that is both ethically and medically sound, and that is culturally competent. Ensure that the strategy employs fair and appropriate utilization of medical resources, and that the approach is practical and feasible within the limits of the medical system at large. Work out a clear and professional way to communicate the proposal to the patient and his family. Attempt to foresee challenges that may arise in conveying or implementing the plan. Determine what follow-up will be necessary to ensure that the chosen strategy remains successful for the patient in the long-term. Reflect on how the knowledge and skills learned from this case can be used to improve the care of patients that may be encountered in future practice.

Afterthoughts

Dennis had an underlying diagnosis of ulcerative colitis, but developed narcotic dependence as a result of his treatment. His increasingly difficult behavior made it both challenging and unpleasant for his physicians to manage his care. For what reasons may a physician decide and justify that they no longer want to care for a patient – where is the ethical line drawn?

Annotated References/Further Information

Code of Medical Ethics of the American Medical Association: Current Opinions with Annotations, 2006–2007 Edition. Council of Ethical and Judicial Affairs. Annotations prepared by the Southern Illinois University Schools of Medicine and Law. The subsection on Termination of the Physician-Patient Relationship (8.115) is relevant to this case.

Snyder L, JD, and Leffler C, JD. Ethics Manual, Fifth Edition. Ethics and Human Rights Committee, American College of Physicians. Annals of Internal Medicine. 142(7):560–582, 5 April 2005. The subsection on Initiating and Discontinuing the Patient-Physician Relationship includes helpful guidance on avoiding patient abandonment.

Public Health Service Act (1975) Title VI. 42 Code of Federal Regulations Part 124.603(b) Emergency Services.

Brennan F et al. Pain management: a fundamental human right. Anesthesia & Analgesia. 105(1):205–221. July 2007.

Ballantyne JC. Opioids for chronic nonterminal pain. Southern Medical Journal. 99(11):1245–1255. November 2006.

Clark JD. Chronic pain prevalence and analgesic prescribing in a general medical population. Journal of Pain and Symptom Management. 23(2):131–137. February 2002.

Katz N and Fanciullo GJ. Role of urine toxicology testing in the management of chronic opioid therapy. The Clinical Journal of Pain. 18:S76–S82.

Dirkzwager AJE and Verhaak PFM. Patients with persistent medically unexplained symptoms in general practice: characteristics and quality of care. BMC Family Practice. 8:33. 31 May 2007.

The Drug and Alcohol Services Information System (DASIS) Report. November 19, 2004 and July 23, 2004. The DASIS Report is published periodically by the Office of Applied Studies, Substance Abuse and Mental Health Services Administration. Acquired from: *http://www. drugabusestatistics.samhsa.gov*. Accessed October 15, 2007.

www.hhs.gov/ocr/hbcsreg.html. Accessed October 15, 2007.

Case 10
When a Patient Makes Questionable Decisions

The Patient

Stephanie D. is a 39-year-old female with a history of intravenous drug abuse and HIV/AIDS, with a most recent CD_4 count of $9/\mu L$, who presented to the Emergency Department with acute onset of upper abdominal pain and bright red hematemesis since the morning of admission. She reported noticing easy bruising and gum bleeding over the past several weeks, but no overt blood loss until the current episode. She also had an increasingly poor appetite and a 20-pound weight loss over the last two months.

On physical exam, she was a cachectic female in no apparent distress. Her physical exam was notable for a mildly decreased blood pressure of 107/51, scattered petechiae on her skin and a moderately tender upper abdomen without rebound or guarding. Laboratory evaluation was remarkable for a decreased white blood cell count of $2,000/\mu L$, a decreased hemoglobin of 6.4 gr/dL and a markedly decreased platelet count of $7,000/\mu L$. Additionally, her blood urea nitrogen was 37 mg/dL and creatinine 1.1 mg/dL, and coagulation studies were within normal limits. She was admitted for further evaluation and treatment of her gastrointestinal bleeding and pancytopenia.

The Ethical Dilemma

The medical team consulted the GI and hematology services, and ordered Stephanie to be *nil per os* (nothing by mouth) and to receive IV fluids, transfusion of red blood cells for her anemia, and platelets in an attempt to stop further blood loss. When the team attempted to obtain informed consent for the transfusion of blood products, Stephanie agreed to transfusion of red blood cells, but refused to accept transfusion of platelets. The team explained to Stephanie that the units of platelets were necessary to help prevent further bleeding, but she still refused to allow them. When pressed for her reasoning behind the decision, she stated "because that's what I want," and wouldn't clarify the matter any further. Although bewildered by her

From: *Evidence-Based Medical Ethics*
By: J.E. Snyder and C.C. Gauthier © Humana Press, Totowa, NJ

decision making on the matter of transfusion, the team determined that Stephanie otherwise had the mental capacity to make decisions regarding her medical care.

Questions for thought and discussion: Can a patient be allowed to make decisions that don't make sense to her medical team? What should she demonstrate to the team to prove her capacity to make these decisions if she otherwise seems "competent"?

Since Stephanie was so ill, and because she continued to have a high risk of bleeding based on her refusal of platelet transfusion, the medical team discussed code status with her. Stephanie told the team that "CPR is okay," and that they could use "one round of medication," but that they could not intubate her or place her on a mechanical ventilator.

Questions for thought and discussion: Can Stephanie continue to dictate her care in ways that seem illogical to her health care providers (e.g., no intubation), especially if those decisions potentially render certain treatment options (e.g., CPR) less effective or ineffective?

The Medicine

Pancytopenia can be due to numerous reasons. When thrombocytopenia is present, data suggests that it can be associated with increased risk for gastrointestinal mucosal bleeding. The resulting anemia from gastrointestinal bleeding may cause increased morbidity and mortality for a patient, although multiple factors such as other underlying disease processes play an important role in this effect. The association between the degree of anemia and patient outcomes has been inadequately studied and there is conflicting evidence about transfusion thresholds for patients. In general many experts recommend basing transfusion decisions for red blood cells on the patient's rate of blood loss, symptomatology, comorbid illnesses, and level of hemodynamic instability. Data for platelet transfusion thresholds is even more limited. Spontaneous bleeding from thrombocytopenia probably won't occur at levels above 10,000/mm^3, although a transfusion threshold of 50,000/mm^3 is often used. Platelet function can also vary by patient, particularly in the setting of certain medications they may be taking, and in chronic liver and kidney disease. Hemostasis, as judged by thorough clinical and laboratory evaluation, may be the best method to determine if platelet transfusion is necessary.

According to American Heart Association (AHA) guidelines for treating cardiac arrest with cardiopulmonary resuscitation (CPR), either bag-mask ventilation or ventilatory support through an advanced airway such as an endotracheal tube, are both adequate techniques for maintaining oxygenation in the patient. There is no prospective data doing a head-to-head comparison between these two methods of ventilatory support in a code situation. Retrospective outcomes data is limited, but shows no advantage to endotracheal intubation. Bag-mask ventilation does increase the likelihood of gastric distention and aspiration, compared to ventilation through an advanced airway. Additionally, the provider must seek to prevent air trapping in the lungs, particularly in patients with chronic obstructive lung disease. If the bag-mask apparatus does not adequately ventilate the patient, then obtaining an advanced airway may be necessary. It is essential that proper technique is used, regardless of the method chosen to ventilate the patient, to ensure maximal oxygen delivery occurs during resuscitation.

Ultimately, if resuscitation is successful in restoring a perfusing rhythm, then having an advanced airway in place will be essential and is considered standard of care. Placement should be done by an experienced individual to minimize the potential for complications. An endotracheal tube helps provide a patent airway, allows for suctioning, protects against aspiration, and allows the patient to be connected to a mechanical ventilator for delivery of controlled tidal volumes and percent inspired oxygen. Additionally, certain emergency medications can be administered through the tube, although this is generally considered not to be a preferred route.

The Law

If Stephanie is able to make her own medical decisions she should be allowed to make decisions that don't make sense to her medical team, within some very important limits. Stephanie has the legal right to accept or refuse medical tests and procedures recommended by her physicians. However, she does not have the legal right to dictate the specifics of her care. If her decisions make treatment options less effective or ineffective, or if her practitioners are not comfortable with her choices, they do not need to provide care in the way she has chosen. If initiating CPR without permission to intubate Stephanie or place her on the ventilator violates the medical team's understanding of good medical practice, they should tell her they are not able to initiate CPR under these conditions.

Stephanie's seemingly random refusal of platelets, when she will accept red blood cells, may raise questions about her capacity to make medical decisions. The medical team was right to explain the importance of the platelets to stop Stephanie's bleeding and to ask Stephanie about her reasons for refusing them. Often eliciting the patient's reasons for a decision can assist the medical team in assessing the patient's capacity. However, it is not legally necessary for a patient to explain the reasons for a particular decision to demonstrate capacity to make that decision.

The capacity to make a specific medical decision is based, at the least, on an understanding of the relevant information and the consequences of accepting or refusing the recommended medical procedures. Most commentators agree that patients may have the capacity to make some medical decisions but not others, based on the complexity of the decision and the complexity of the relevant information. The medical team should determine whether or not Stephanie has the capacity to make these particular decisions by ensuring she understands the information they have provided and the consequences (e.g., refusing the platelets may be life-threatening).

If the medical team cannot determine the rationale behind Stephanie's random choices and her refusal to explain her reasons, an independent assessment of her decision-making capacity may be needed. In that case, the medical team could ask for a psychiatric consult to better determine her decision-making capacity.

The Ethics

The recommendations of the medical team were based on the *Principles of Beneficence* and *Non-Maleficence*. The transfusion of red blood cells was needed to promote Stephanie's health by treating her anemia. The transfusion of platelets was needed to prevent further blood loss. The medical team also wanted to avoid doing harm to Stephanie by providing less effective treatment in the form of a blood transfusion without platelets and attempting CPR without the options of intubation and mechanical ventilation.

According to the *Principle of Respect for Autonomy*, capable patients should be allowed to make decisions about their own medical care. Patients need to be given relevant information about the treatment options recommended by practitioners in language they can understand. Patients may then accept or refuse the recommended medical procedures. Stephanie was given information about the transfusion of blood products. When she refused the platelets, the medical team explained how important they were to stop further bleeding. When she continued to refuse and could not or would not explain her reasons, her capacity to make these decisions was questioned.

To prove that she is capable of making these decisions, Stephanie needs to show she understands the recommended procedures and the consequences of refusing them. If Stephanie is determined to be capable of decision making in this situation, the medical team should not impose treatment on her that she has refused. In particular, they should not transfuse her with platelets.

However, this does not mean that she can dictate the details of her medical care. In regard to the code status question, Stephanie thinks she can accept CPR and "one round of medication," but refuse intubation and mechanical ventilation. If the medical team believes that CPR with these restrictions would be ineffective, they probably should not have asked her about these specific aspects of CPR. Then they could explain to Stephanie that they are unwilling to attempt CPR with these limitations because this violates their conception of good medical practice and accepted standards of patient care.

It might also be helpful to inquire about Stephanie's family support or close friends that she might want involved in her medical care and decision making. If there were other people Stephanie felt comfortable talking to about these decisions, it might help clarify to the medical team the reasons for her choices. This approach would be supported by the *Principle of Respect for Dignity*.

The Formulation

Now that the evidence-based medicine, legal precedent, and relevant ethical principles for this case have been reviewed, formulate a strategy to address the ethical conflicts in this case. If necessary, perform additional research into local and state laws and hospital regulations. Consider delving further into the background medical literature to assist with making sound therapeutic decisions. Devise a treatment approach that addresses the needs of the patient and her family, that is both ethically and medically sound, and that is culturally competent. Ensure that the strategy employs fair and appropriate utilization of medical resources, and that the approach is practical and feasible within the limits of the medical system at large. Work out a clear and professional way to communicate the proposal to the patient and her family. Attempt to foresee challenges that may arise in conveying or implementing the plan. Determine what follow-up will be necessary to ensure that the chosen strategy remains successful for the patient in the long-term. Reflect on how the knowledge and skills learned from this case can be used to improve the care of patients that may be encountered in future practice.

Afterthoughts

Stephanie dictated aspects of her care that were medically illogical. This created an ethical dilemma for her health care providers who were attempting to reverse her poor condition, but were limited in their therapeutic options due to her questionable decision making and lack of consent. What qualifies a patient's choice as based in a belief system that should be respected versus one that is based on muddled logic? Is capacity all-or-nothing with regard to decision making, or can patients truly have the capacity to make some decisions and not others?

Annotated References/Further Information

Code of Medical Ethics of the American Medical Association: Current Opinions with Annotations, 2006–2007 Edition. Council on Ethical and Judicial Affairs. Annotations prepared by the Southern Illinois University Schools of Medicine and Law. The subsection on Informed Consent (8.08) is relevant to this case.

Snyder L, JD, and Leffler C, JD. Ethics Manual, Fifth Edition. Ethics and Human Rights Committee, American College of Physicians. Annals of Internal Medicine. 142(7):560–582, 5 April 2005. In the subsection on Informed Consent the authors explain the consensus view that patients may have the capacity to make some decisions, but not others.

Beauchamp TL and Childress JF. Principles of Biomedical Ethics, Fifth Edition, Oxford University Press, 2001. In a section on The Capacity for Autonomous Choice (pp. 69–77) the authors examine the various standards and requirements for competence and decision-making capacity.

Maltz GS et al. Hematologic management of gastrointestinal bleeding. Gastroenterology Clinics of North America. 29(1):169–186. March 2000.

Wallace JL et al. Platelets accelerate gastric ulcer healing through presentation of vascular endothelial growth factor. British Journal of Pharmacology. 148(3):274–278.

2005 American Heart Association Guidelines for Cardiopulmonary Resuscitation and Emergency Cardiovascular Care. Part 7.1: Adjuncts for Airway Control and Ventilation. *Circulation*. 2005;112:IV-51–IV-57.

Case 11
When a Partner Is Excluded

The Patient

Doug M. is a 34-year-old male with a history of HIV infection who presented to the Emergency Department with a two-month history of cough productive of scant yellow sputum, malaise, subjective fevers, and a 15-pound weight loss. In the two weeks prior to admission he developed progressively worsening shortness of breath and weakness, prompting him to seek medical attention. Doug had been diagnosed with HIV five years prior to admission, but had stopped taking his antiretroviral medications and had not followed up with his primary physician for the past two years due to the financial constraints of changing jobs and losing his health insurance. At his last visit to his physician two years ago, his CD_4 count was documented to be 201/µL. Doug denied the use of tobacco, alcohol, or illicit drugs. He was currently working at a day care center and owned a home with his longtime partner, Justin. Justin was HIV-negative; Doug had acquired HIV through a prior sexual contact. Doug's family history was noncontributory, although his mother had passed away several years prior in a motor vehicle accident.

On physical exam Doug had a low-grade fever, was mildly tachycardic and had an oxygen saturation of 86 percent on room air. Chest auscultation revealed decreased breath sounds and scattered wheezes. Initial laboratory evaluation was unremarkable except for a relative lymphopenia present on his complete blood count. Chest radiography was consistent with diffuse, bilateral interstitial infiltrates. Sputum was collected and sent for gram stain, acid-fast smear for Mycobacteria, and methenamine silver stain for *Pneumocystis jirovecii*, all of which had unrevealing results. Doug was admitted to the inpatient medical team and started on empiric treatment for both community-acquired and *Pneumocystis jirovecii* pneumonia (PJP), the latter based on strong clinical suspicion for the diagnosis. Subsequently, his CD_4 count was reported to be 12/µL. The infectious disease team was consulted and agreed with the primary team's antibiotic choices. They also recommended that Doug be restarted on antiretroviral medications.

From: *Evidence-Based Medical Ethics*
By: J.E. Snyder and C.C. Gauthier © Humana Press, Totowa, NJ

The Ethical Dilemma

Over the first several days of admission Doug's respiratory status worsened. Bronchoscopy revealed the presence of pneumocytes and *Pneumocystis jirovecii* organisms. Doug's oxygen needs continued to increase, and he was transferred to the medical intensive care unit where he was subsequently intubated and placed on a mechanical ventilator. Shortly thereafter, chest radiography was consistent with a diagnosis of acute respiratory distress syndrome (ARDS). Doug also developed acute renal failure, a coagulopathy, and anasarca. As the prognosis became increasingly grim, the medical team continued to inform Doug's partner, Justin, of his condition. Doug and Justin, although partnered for several years, had never filled out legal documentation such as a Health Care Power of Attorney and they lived in a state where same-sex relationships had no legal recognition. By current state law Doug's father was his next of kin and, hence, his surrogate decision maker. Justin reported to the medical team that Doug visited his father frequently, but had never divulged his homosexuality or HIV diagnosis to his father or other family members. Therefore, Justin had never met Doug's father. He felt that Doug had decided to keep his relationship with another man secret out of concern that his father might not have an accepting attitude towards it.

> **Questions for thought and discussion:** Does the medical team need to contact Doug's father about his condition? How much information should be disclosed to Doug's father about his health and life outside the hospital?

The medical team contacted Doug's father and asked him to come to the ICU to see Doug and to have a meeting about his condition. He first visited Doug's room to spend some time with his son, and then met with the medical team. He started the family meeting, unprompted, by asking the medical team "does my son have AIDS?" He went on to explain that he had suspected his son was gay for years, although they had never discussed it previously, and that he "figured he was dying of AIDS" as a result.

> **Questions for thought and discussion:** What, if any, is the role of the medical team to dispel the perception Doug's father has that being gay is the same as having AIDS? Do his father's perceptions hinder his ability to act as a surrogate decision maker for Doug?

Concerned that Doug was, in fact, dying and that informed end-of-life decisions would need to be made regarding issues such as resuscitation status, the team decided to disclose the entire truth about Doug's sexuality, diagnosis, and condition to his father. The team perceived that Doug's father was able to adequately set aside his personal beliefs and feelings, and make competent choices regarding Doug's treatment. As it seemed that Doug's care was futile despite optimal medical therapy and respiratory support, the team proposed terminal extubation and a shift toward comfort care only to Doug's father. He agreed with the concept, but placed the stipulation on the decision that "that man," referring to Justin, not be allowed to visit Doug's room anymore.

Questions for thought and discussion: Can Doug's father exclude Justin from the room at the end of Doug's life? Does Justin have any rights in this situation?

Question for thought and discussion: Would it be ethical or helpful to share with Doug's father that Justin is HIV-negative and that Doug did not acquire HIV from him?

The Medicine

According to 2005 data from the Centers for Disease Control and Prevention (CDC), the most common route of transmission in AIDS cases to date is via sexual contact in men who have sex with men (MSM). This accounts for about 59 percent of cases in adult and adolescent males. Approximately 22 percent of cases in males are attributable solely to injection drug use (IDU), 8 percent of cases are in MSM with a history of IDU and about 8 percent is due to high risk heterosexual contact. In adult and adolescent females approximately 40 percent of cases are due to transmission via IDU and 56 percent are via high risk heterosexual contact. The CDC estimates that around one-quarter of persons infected with HIV in the United States are unaware of their status.

In epidemiologic studies of HIV transmission, MSM refers to any man who has sex with a man. MSM is a grouping which includes persons who self-identify as gay, bisexual, or heterosexual, and hence includes men that may also have sexual relations with women. Sociocultural stigmatization of homosexuality may prevent some MSM from self-identifying as gay or bisexual. Actual HIV prevalence data among men who self-identify as homosexual is not readily available and it seems that high transmission risk seems to be correlated with behaviors more complex than just sexual orientation. In one study of 4,998 MSM who were HIV-negative, Bartholow, et al. determined that older men with multiple partners, older men using nitrate inhalants, and young men using amphetamines and hallucinogenic drugs were the most likely MSM subgroups to become infected with HIV. In a report from the Centers for Disease Control and Prevention (CDC) that focused on the risk for young men (aged 15 to 29) of acquiring HIV and other STDs, those MSM who were "non-disclosers" (those who did not publicly reveal themselves to be an MSM, commonly referred to as "closeted") were also less likely to self-identify as homosexual. Of note, "non-disclosing" black men in the study were more likely than their "out" counterparts (those who publicly acknowledge their bisexual or homosexual orientation) to not use HIV testing services, to be unaware of their positive HIV status, and were more likely to have recently engaged in heterosexual intercourse.

According to United States data from the CDC, 41,897 persons were diagnosed with AIDS in 2005 and 17,011 people with AIDS died in 2005. And cumulatively, as of 2005, there have been 984,155 diagnoses of AIDS in the United States and 550,394 deaths. Despite the advent of antiretroviral therapies for HIV infection, *Pneumocystis jirovecii* pneumonia (PCP) is still the most common opportunistic infection in HIV-infected persons, and is a major cause of mortality in patients with AIDS. Chemoprophylaxis against *Pneumocystis* infection with a medication such as trimethoprim-sulfamethoxazole (TMP-SMX) is indicated in all HIV-infected individuals with CD_4^+-lymphocyte counts below 200 per milliliter. Many cases of PCP still occur due to the high number of persons unaware of their positive HIV status. Additional cases occur in others who are poorly- or non-adherent with antiretroviral therapy or PCP chemoprophylaxis. Additionally, there is occasionally infection with a *Pneumocystis* strain that develops resistance to TMP-SMX.

Mortality from PCP is reduced in patients with $P_aO_2 < 70mmHg$ (or an alveolar-arterial gradient of $>35\,mmHg$ on room air) when corticosteroids are added to the treatment regimen. In a study by Kumar and Krieger the mortality of patients with PCP who required mechanical ventilation was 81 percent, with increasing mortality rates in subgroups of patients whose CD_4^+-lymphocyte counts approached zero per milliliter. The APACHE (acute physiology and chronic health evaluation) scoring system, used to predict mortality in critically ill patients, may underestimate mortality in patients with PCP requiring mechanical ventilation. Azoulay, et al. found that determinants of increased 90-day mortality in patients with PCP included hypoxemia on admission (P_aO_2 of 60mmHg on room air), elevated lactate dehydrogenase levels (>1,000 I.U.), low serum albumin (<30 g/L), neutrophilia in bronchoalveolar lavage fluid (>10%), development of a pneumothorax, acquisition of a nosocomial infection, and the need for mechanical ventilation. When treatment for PCP is not effective or initiated with delay, PCP can progress to acute respiratory distress syndrome (ARDS) which, on its own, carries a 34 to 58 percent mortality rate. Factors that reliably predict futility of care for PCP do not yet exist.

The Law

When a patient has not designated a surrogate decision maker for themselves via a written document such as a Health Care Power of Attorney, and they do not have the capacity to make their own health care decisions, then the surrogate is generally decided by state law. A number of states have designated means to determine, by level of priority, who the responsible surrogate should be in such situations (see Table 11.1), and health care practitioners should familiarize themselves with the law in their state of practice before assigning a surrogate to this role. If a person who is legally authorized as a surrogate decision maker is unable or declines to accept the responsibility of this role, then the person of next highest priority is given authority.

The central theme of these laws is that a patient's closest biological or legally-recognized family has preferred status. The order of priority is most commonly a court-appointed legal guardian, the patient's legal spouse, the adult children of the

Table 11.1 The hierarchy of possible surrogate decision makers for a patient*

The health care agent listed on a Power of Attorney for Health Care
↓
The patient's legal guardian
↓
The patient's legal spouse
↓
The adult child(ren) of the patient
↓
The parent(s) of the patient
↓
The patient's adult sibling(s)
↓

Other persons (e.g., relatives, friends) well-acquainted with the patient, who have had reasonable contact with them and demonstrate a good understanding of their beliefs and wishes

*In 32 states. Please refer to individual state laws for proper determination of the proper surrogate decision maker as this varies from state to state and is an area of constantly evolving legislation

patient, the patient's parents, the patient's adult siblings, or another relative or friend of the patient who has had reasonable contact with them and demonstrates a good understanding of their beliefs and wishes. Since Doug and Justin live in a state where the relationship between same-sex couples is not legally recognized, Doug's father officially has the right to be his surrogate decision maker. Doug's physicians need to decide if Doug's father has the capacity to fulfill this role. If he does not, then an alternative person must be named. By most existing state laws, Justin would be considered to be "another friend or relative," and if Doug has adult siblings or children, they would also have higher authority than Justin by law. Although it did not happen in this case, Doug's father could have declined the role of surrogate to let Justin assume this role.

Although several states have passed legislation giving same-sex partners the legal rights afforded to married heterosexual couples (see Table 11.2), the majority of states do not offer such legal protection and most have laws or constitutional amendments specifically barring marriage rights from same-sex couples. In order for same-sex partners to serve as each others' surrogate decision makers in most states, they need to be named as the health care agent on a legal document, such as a Health Care Power of Attorney, with a document completed by each partner.

Doug's father has chosen to exclude Justin from his dying partner's hospital room. As the surrogate decision maker for Doug, his father can legally control visitation to Doug's room. However, one may question if the father is acting in Doug's best interest by doing so, since Justin had Doug's permission to visit him earlier in the hospital course. This requires further investigation by the medical team to assess Doug's father's capacity to make good health care decisions for Doug.

Sharing Justin's HIV status with Doug's father is strictly forbidden by laws pertaining to confidentiality. The Health Insurance Portability and Accountability Act of 1996 (HIPAA) gives individuals the right to limit access to their protected health information (PHI) to persons of their choosing and certain persons directly involved in their own medical care. If Justin consents to disclose his HIV status to others, and the medical team feels that this is essential information that could

Table 11.2 Legal recognition of same-sex relationships in the United States

States that issue marriage licenses to same-sex couples
    Massachusetts, 2004
States legally providing same-sex couples with equivalent to spousal rights in-state
    Vermont (offers civil unions since 2001)
    Connecticut (offers civil unions since 2005)
    California (offers domestic partnerships since 2006, marriage licenses approved for 2008)
    New Hampshire (offers civil unions since 2008)
    New Jersey (offers civil unions since 2007)
    Oregon (offers domestic partnerships since 2008)
States legally providing same-sex couples with some spousal right equivalents in-state
    Hawaii (offers reciprocal beneficiaries since 1997)
    Maine (offers domestic partnerships since 2004)
    Washington (offers domestic partnerships since 2007)
    District of Columbia (offers domestic partnerships since 1992, but not effective
       until 2002 due to Congress prohibiting law implementation)

Source adapted from: The Human Rights Campaign (*www.hrc.org*). Accessed
October 15, 2007.

impact Doug's care, only then can this information be provided to Doug's father.
Whether this disclosure would even impact Doug's care is debatable.

The Ethics

According to the federal Patient Self-Determination Act (1990), when capable
patients are admitted to the hospital, they should be asked whether or not they have
an advance directive and this should be documented in their medical chart. Doug
didn't have an advance directive. Perhaps, if Doug was conscious and capable when
he was first admitted, particularly given his diagnosis, someone on the medical team
should have asked him who he wanted to make decisions for him when he no longer
could. If Doug had named Justin, the answer could have been documented in Doug's
medical chart and this note could have served as evidence of his wishes.

This may have legally allowed Justin to become Doug's surrogate decision
maker. In most states, statutory advance directives are not meant to be the only way
patients can make their wishes known. Written evidence of the patient's wishes in
other forms is also acceptable. The medical team must document Doug's verbal
wishes in his chart. In this case, the medical team should not contact Doug's father
or share any information about his medical condition, without Doug's permission,
based on medical confidentiality.

If Doug lost consciousness before he could name Justin as the person he wanted
to make decisions for him, then Doug's father must be contacted. At this point,
someone needs to make decisions about Doug's medical treatment and his father is,
in most states, the one legally authorized to do so. When he is contacted, Doug's
father should be given complete information about Doug's diagnosis, his present
condition and his prognosis, but not about his sexuality.

Doug's father does not need to be given any information about Doug's sexuality or his relationship with Justin. Doug did not want his father to have this information and, more importantly, it is not necessary for any medical decisions that must be made for Doug. Based on the *Principle of Respect for Dignity*, Doug's confidentiality must be protected as far as possible, consistent with good medical care. Similarly, the *Principle of Veracity* does not require that Doug's father be given this information because he doesn't need to know about Doug's sexuality or his relationship to make medical decisions for him.

The medical team could talk about the difference between being gay and having AIDS with Doug's father, if they thought this would help the father accept Doug's diagnosis. On the other hand, they need to remember that Doug – not his father – is their patient and their actions should be focused on Doug's best interests and not those of his father. With this in mind, they should not be discussing Doug's sexuality with his father, since this is totally irrelevant to his medical treatment and would violate his confidentiality.

However it is handled, the father's misconception does not disqualify him as Doug's surrogate decision maker. The relationship between sexuality and AIDS is not relevant to treatment decisions that need to be made now for Doug. If Doug's father is able to understand the information provided to him about Doug's condition, his prognosis, the recommendations of the medical team, and the consequences of comfort care for Doug, he should be able to serve as Doug's surrogate decision maker.

Using the *Principles of Beneficence* and *Non-Maleficence*, everyone involved in this situation appears to be considering what is in the patient's best interests in terms of end-of-life medical treatment. The medical team has recommended that Doug be extubated and given comfort care based on promoting good and preventing/avoiding harm. Doug's father, as his surrogate decision maker, has agreed and Doug's partner, Justin, has not raised any objections to this decision. There is agreement, then, that a natural death, without life-prolonging interventions, would be best for Doug.

The *Principle of Respect for Autonomy* is going to be very difficult to apply in this case. No one asked Doug during his last illness or at his admission to the hospital what he wanted in terms of end-of-life treatment, or who he wanted to make treatment decisions if he no longer could. This makes it impossible to know for sure what Doug would want. Based on the *Principle of Respect for Dignity*, however, the medical team has considered Doug's relationship with Justin and has included Justin in conversations about Doug's condition and care. Justin has also been able to provide some insight into Doug's relationship with his father.

Ethically, Justin would be the best choice as surrogate decision maker for Doug, based on the *Principle of Respect for Dignity*, considering the importance of their relationship and their emotional connection. This is a case where ethics and the law disagree, since Doug's father is going to be his legally-authorized surrogate, in most states. Fortunately, Doug's father has agreed to the recommendation of the medical team and there seems to be no objection from Justin.

Although Justin really has no legal rights in this situation, the father's demand that Justin not be allowed in Doug's room is cruel, unnecessary, and not in Doug's best interests. It ignores Doug's human dignity, in particular his emotions, relation-ships, and likely goals for the end of his life. It is also questionable whether Doug's father has the right, even as surrogate decision maker, to make this demand. It has nothing to do with Doug's medical treatment. The medical team could explain that Justin has been in the room with Doug from the beginning and who is in the room is not a treatment issue so this is not up to Doug's father.

If it was meant as a threat – "I'll agree to what you recommend as long as you keep that man away" – it is a weak one. It is very unlikely that Doug's father is going to demand CPR and life-prolonging procedures for Doug, just because Justin is allowed to stay in Doug's room.

On the other hand, the father's negative reaction may be in part a result of the lack of time he has had to accept both Doug's illness and his relationship with Justin. The father did not know that his son had HIV or that it had progressed to AIDS. He did not know that Doug had a long-term partnership with Justin. His emotional response at suddenly learning that his son is dying will naturally include anger and this is very likely to be directed toward the person who has been the closest to his son all the time he has been sick. He may be feeling that Doug should have let his father care for him during his illness, rather than some-one who is, to the father, a "stranger."

Even while the medical team is making an effort to understand the father's reac-tion and explaining their position about Justin being in Doug's room, they should not try to ease the situation by telling Doug's father that Justin is HIV-negative. This would violate Justin's confidentiality, no matter how the medical team learned this information. As Doug's partner, Justin deserves to be treated with respect for his dignity as well.

The Formulation

Now that the evidence-based medicine, legal precedent, and relevant ethical principles for this case have been reviewed, formulate a strategy to address the ethical conflicts in this case. If necessary, perform additional research into local and state laws and hospital regulations. Consider delving further into the background medical literature to assist with making sound therapeutic decisions. Devise a treatment approach that addresses the needs of the patient and his family, that is both ethically and medically sound, and that is culturally competent. Ensure that the strategy employs fair and appropriate utilization of medical resources, and that the approach is practical and feasible within the limits of the medical system at large. Work out a clear and professional way to communicate the proposal to the patient and his family. Attempt to foresee challenges that may arise in conveying or implementing the plan. Determine what follow-up will be necessary to ensure that the chosen strategy remains successful for the patient in the long-term. Reflect on how the knowledge and skills learned from this case can be used to improve the care of patients that may be encountered in future practice.

Afterthoughts

A patient's same-sex partner was excluded from participating in health care decisions due to a lack of legal rights. The medical team had to determine if the legal surrogate decision maker had the capacity to make appropriate health care decisions for the patient, given his own personal belief system. The case emphasizes the importance of legal documentation so patients can designate the person they choose to assume the role of surrogate, particularly if the law would assign someone else by default.

Could Doug's medical team have arranged for him to complete legal documentation early in his hospitalization, prior to his deterioration, assigning Justin as his surrogate? Should health care practitioners ask all patients admitted to the hospital, regardless of age or diagnosis, who their choice for a surrogate would be?

Should a health care provider ever divulge their own sexual orientation to a patient or their family? Would this information ever be relevant to the care of a patient? Could this information ever be helpful in establishing a stronger therapeutic bond with a patient or their loved ones? Could it ever be detrimental? Should a provider's other cultural background information (religious affiliation, for example) ever be shared in a therapeutic relationship with a patient?

Annotated References/Further Information

Code of Medical Ethics of the American Medical Association: Current Opinions with Annotations, 2006–2007 Edition. Council on Ethical and Judicial Affairs. Annotations prepared by the Southern Illinois University Schools of Medicine and Law.

Snyder L, JD, and Leffler C, JD. Ethics Manual, Fifth Edition. Ethics and Human Rights Committee, American College of Physicians. Annals of Internal Medicine. 142(7):560–582, 5 April 2005.

http://www.hrc.org. Accessed October 15, 2007.

Decisions near the end of life. Council on Ethical and Judicial Affairs, American Medical Association. JAMA. 1992;267:2229–2233.

Patient Self Determination Act (1990). 42 U.S.C. 1395 cc (a). Subpart E. Section 4751. The Patient Self-Determination Act can be found at *www.dgcenter.org/acp/pdf/psda.pdf*. Accessed October 15, 2007.

http://www.cdc.gov/hiv/topics/surveillance/basic.htm#hivest. Accessed October 15, 2007.

http://www.cdc.gov/hiv/topics/msm/resources/factsheets/msm.htm. Accessed October 15, 2007.

Prevention toolkit. CAPS Fact Sheet – What are men who have sex with men (MSM)'s HIV prevention needs? Revised 12/2000. Center for AIDS Prevention Studies (CAPS) at the University of California San Francisco. Acquired from: *http://www.caps.ucsf.edu/pubs/FS/MSMrev.php*. Accessed October 15, 2007.

Glynn M and Rhodes P. Estimated HIV prevalence in the United States at the end of 2003. National HIV Prevention Conference; June 2005; Atlanta. Abstract T1-B1101.

CDC. HIV/AIDS among racial/ethnic minority men who have sex with men – United States, 1989–1998. MMWR. 2000;49:4–11.

Bartholow BN et al. Demographic and behavioral contextual risk groups among men who have sex with men participating in a phase 3 HIV vaccine efficacy trial: implications for HIV prevention and behavioral/biomedical intervention trials. Journal of Acquired Immune Deficiency Syndromes. 2006;43:594–602.

HIV/STD risks in young men who have sex with men who do not disclose their sexual orientation — Six U.S. Cities, 1994—2000. CDC's Morbidity and Mortality Weekly Report. 2003;52(5): 81–86. Acquired from: *http://www.cdc.gov/mmwr/preview/mmwrhtml/mm5205a2.htm*. Accessed October 15, 2007.

Ashley EA et al. Human immunodeficiency virus and respiratory infection. Current Opinion in Pulmonary Medicine. 2000;6:240–245.

Rosen MJ and Narasimhan M. Critical care of immunocompromised patients: human immunodeficiency virus. Critical Care Medicine. 2006;34(9 Suppl.):S245–S250.

Azoulay E et al. AIDS-related *Pneumocystis carinii* pneumonia in the era of adjunctive steroids. Implication of BAL Neutrophilia. American Journal of Respiratory and Critical Care Medicine. 1999;160:493–499.

Kumar SD and Krieger BP. CD4 lymphocyte counts and mortality in AIDS patients requiring mechanical ventilator support due to PCP. Chest. 113:430–433. 1998.

Briel M et al. Adjunctive corticosteroids for Pneumocystis jiroveci pneumonia in patients with HIV-infection. Cochrane Database of Systematic Reviews. Date of Most Recent Update: 24-May-2006. Date of Most Recent Substantive Update: 24-May-2006.

MacCullum NS and Evans TW. Epidemiology of acute lung injury. Current Opinion in Critical Care. 2005;11:43–49.

Kaplan JE et al. Epidemiology of human immunodeficiency virus-associated opportunistic infections in the United States in the era of highly active antiretroviral therapy. Clinical Infectious Diseases. 2000;30 Suppl 1:S5–14.

Randall CJ et al. Improvements in outcomes of acute respiratory failure for patients with human immunodeficiency virus-related Pneumocystis carinii pneumonia. American Journal of Respiratory and Critical Care Medicine. 2000;162:393–8.

Morris A et al. Current epidemiology of *Pneumocystis* pneumonia. Emerg Infect Dis [serial on the Internet]. 2004 Oct. Acquired from: *http://www.cdc.gov/ncidod/EID/vol10no10/03-0985.htm*. Accessed October 15, 2007.

Cases of HIV infection and AIDS in the United States and dependent areas, 2005. HIV/AIDS Surveillance Report, Volume 17, Revised Edition, June 2007. Acquired from: *http://www.cdc.gov/hiv/topics/surveillance/resources/reports/2005report/default.htm*. Accessed October 15, 2007.

Curtis JR et al. Improvements in outcomes of acute respiratory failure for patients with human immunodeficiency virus-related *Pneumocystis carinii* pneumonia. American Journal of Respiratory and Critical Care Medicine. 2000;162:393–398.

Case 12
When a Diagnosis Is Reportable

The Patient

Scott O. is a 37-year-old male with no significant past medical history who presented to the Emergency Department with a one week history of worsening dysphagia and odynophagia. He reported he was usually in good health until the current symptoms began. Scott said that he had been unable to swallow solid foods over the past week, largely due to increasing pain on swallowing, and was now unable to swallow liquids as well. Although he stated that his appetite was unchanged, he estimated a five-pound weight loss over the last week due to poor oral intake. He otherwise denied fever, respiratory or gastrointestinal symptoms, or any other complaints. He takes no medications at home except for rare acetaminophen. He is married, has a two-year-old son, and works as an attorney. He denied use of tobacco, alcohol, or illicit drugs. Family history was noncontributory.

On physical exam, he was afebrile and his vital signs were within normal limits. Cardiovascular, respiratory, abdominal, extremity, and neurologic exam were unremarkable. There was some mild oropharyngeal thrush, palpable bilateral inguinal lymphadenopathy, and a 1 cm painless ulceration on the penile shaft. Laboratory evaluation was unremarkable.

The Ethical Dilemma

Scott was admitted for further evaluation and treatment by the medical team. His wife was asked to step out of the room while the medical resident asked Scott some further questions. The resident told Scott that she was concerned that he had primary syphilis based on his ulceration and adenopathy and stated he should be tested for that as well as for HIV, given his thrush. When she pressed Scott for information about his sexual history he initially hesitated, but then acknowledged that he had in fact had sexual relations with a female prostitute, "but only once." He consented to tests for HIV-1 antibodies and syphilis, but begged the resident "please don't tell my wife." Both tests came back positive shortly thereafter. When Scott learned the news, he admitted to the medical resident that he has also had sexual relations with a woman at his office. He again implored that his test results not be shared with his wife or his other sexual partners.

> **Questions for thought and discussion:** What are the obligations of the medical resident for reporting the positive HIV and syphilis test results? Who do these results get reported to? Who notifies the sexual partners of patients who test positive for sexually transmitted infections?

> **Question for thought and discussion:** When should patient confidentiality be broken, even if it is against the patient's wishes?

The Medicine

The advent of antiretroviral therapy has led to improved survival rates for those infected with HIV. Despite national campaigns designed to encourage people to get tested for HIV, around one-quarter of those infected are unaware of their diagnosis and many persons are diagnosed at an advanced stage of disease. In a 2005 report from the Centers for Disease Control and Prevention (CDC) it was estimated that 39 percent of patients diagnosed with HIV are subsequently diagnosed with AIDS in less than 12 months. Among AIDS-defining illnesses (see Table 12.1), esophageal Candidiasis is one of relatively high incidence. Identifying patients with HIV infection prior to their developing an HIV-related illness would be beneficial and

Table 12.1 The most common AIDS-defining illnesses, from highest to lowest incidence

Pneumocystis jirovecii pneumonia
Esophageal candidiasis
Kaposi sarcoma
Mycobacterium avium-intracellulare complex
Cytomegalovirus retinitis
Other cytomegalovirus disease
Recurrent pneumonia
Cryptosporidiosis
Lymphoma
HIV wasting syndrome
Toxoplasmosis
HIV encephalopathy
Cryptococcosis
Other AIDS-defining diagnoses (includes histoplasmosis, isosporiasis, progressive multifocal leukoencephalopathy, and *Salmonella* septicemia)
Pulmonary tuberculosis
Recurrent herpes simplex
Extrapulmonary tuberculosis

Source adapted from: Mocroft A et al. The incidence of AIDS-defining illnesses in 4,883 patients with human immunodeficiency virus infection. Archives of Internal Medicine. 158:491–497; 9 March 1998.

allow antiretroviral treatment to be initiated earlier. One way in which early diagnosis of HIV infection may be better achieved is through more aggressive screening at the primary care provider level. In one study by Gao, et al. it was found that only 28 percent of the over 3,000 patients surveyed were asked about sexually transmitted infections (STIs) at their last routine health visit.

From 1990 to 2000, rates of primary and secondary syphilis declined a sharp 89.7 percent; however, a resurgence has been seen in the subsequent years. The greatest increases have been in men, particularly those having sex with other men (MSM). According to one study by Ciesielski and Boghani, MSM who have syphilis are 10 times more likely than heterosexual males with syphilis to be coinfected with HIV. The chancres caused by syphilis disrupt the natural barriers against HIV infection and increase susceptibility to HIV infection by up to five times. STIs such as syphilis are indicative of risk-taking behavior that is associated with increased transmission of HIV.

Based on 2005 data, the CDC reports that, in adult and adolescent males, AIDS is transmitted via sexual contact in men who have sex with men (MSM) in about 59 percent of cases. About 8 percent is due to high risk heterosexual contact. Although Scott is married to a woman and claims to have had high risk heterosexual encounters, he should be specifically questioned about having past sexual relations with other men. MSM is a grouping which includes persons who self-identify as gay, bisexual, or heterosexual and, hence, includes men that may also have sexual relations with women. Sociocultural stigmatization of homosexuality, particularly in certain minority groups, may prevent some MSM from self-identifying as gay or bisexual. Some men who proclaim themselves to be heterosexual may admit to being "on the down low," an expression referring to those men who occasionally also have sex with other men. Nonjudgmentally asking a patient "do you have sex with men, women, or both?" is probably the best way to obtain such sexual history information during the medical interview.

The Law

By law licensed physicians and other health care providers must report cases, or even suspected cases, of certain medical conditions and communicable diseases that they come upon in practice to their local health department (for those in North Carolina, please see Table 12.2). Cases must be reported in a timely fashion to help prevent epidemic spread of infectious diseases or agents of biological warfare. The list of reportable diseases, mechanism for reporting, and timing of the filed report are established by state and local public health departments, in association with the United States Department of Health and Human Services (HHS). Health care practitioners are encouraged to familiarize themselves with the individual laws and guidelines in their state(s) of practice.

In many states, once the HHS is notified of a reportable disease such as confirmed HIV infection, a health investigator representing that state's HHS will become

Table 12.2 Reportable communicable diseases in North Carolina (may vary by state)

AIDS
Anthrax
Botulism
Brucellosis
Campylobacter infection
Chancroid
Chlamydia
Cholera
Cryptosporidiosis
Cyclosporiasis
Dengue
Diphtheria
E. coli, Shiga Toxin-producing infection (including *E.coli* O157:H7)
Ehrlichiosis, granulocytic
Ehrlichiosis, monocytic (E. chaffeensis)
Encephalitis, Arboviral (CAL, EEE, WNV, other)
Enterococci, Vancomycin-resistant ("VRE"), from normally sterile site
Foodborne Disease (C. perfringens, Staphylococcal, Other/Unknown)
Gonorrhea, all sites
Granuloma Inguinale
Hantavirus infection
Hemolytic Uremic Syndrome
Hemophilus influenzae, invasive disease
Hepatitis A
Hepatitis B, Acute
Hepatitis B, Carrier
Hepatitis B, Perinatal
Hepatitis C, Acute
HIV infection
Legionellosis
Leptospirosis
Listeriosis
Lyme Disease
Lymphogranuloma Venereum
Malaria
Measles
Meningitis, Pneumococcal
Meningococcal Disease
Monkeypox
Mumps
Nongonococcal Urethritis (NGU), other than lab-confirmed Chlamydia
Plague
Pelvic Inflammatory Disease
Polio, paralytic
Psittacosis
Q Fever
Rabies, Human
Rocky Mountain Spotted Fever
Rubella
Rubella, Congenital Syndrome
Salmonellosis
S.A.R.S. (Coronavirus Infection)
Shigellosis

(continued)

Table 12.2 (continued)

Smallpox
Streptococcal infection, Group A, invasive disease
Syphilis, all stages
Tetanus
Toxic Shock Syndrome
Toxic Shock Syndrome, Streptococcal
Toxoplasmosis, Congenital
Transmissible Spongiform Encephalopathies (CJD/vCJD)
Trichinosis
Tuberculosis
Tularemia
Typhoid, Acute
Typhoid Carrier
Typhus, Epidemic (louse-borne)
Vaccinia
Vibrio infection, Other
Vibrio vulnificus infection
Viral Hemorrhagic Fever
Whooping Cough (Pertussis)
Yellow Fever

Source adapted from: North Carolina Communicable Disease Report Card. Acquired from: *http://www.epi.state.nc.us/epi/gcdc/manual/CDReportCard.pdf.* Accessed October 15, 2007.

involved in the case. The investigator will meet with the patient and encourage them to seek medical attention and to use abstinence or barrier methods to help prevent spread of the virus. They will also attempt to determine who the patient may have, through sufficient contact, exposed to the disease. Although they are strongly encouraged to do so, the patient is not required by law to divulge all sexual contacts to the investigator. If a patient has not personally notified their partners of their risk for HIV infection, the investigator may then contact all divulged sexual partners for the need to pursue testing and potential treatment for HIV. This is done in a way that does not disclose who may have exposed them to the virus.

Under the Health Insurance Portability and Accountability Act of 1996 (HIPAA), patient confidentiality laws mandate that Scott's diagnoses of HIV and syphilis must be kept confidential, per his request, from all people who are not at risk for contracting these diseases from him. However, when a patient's known disease presents an imminent threat to the health or life of others, then confidentiality must be broken. Similar to the precedent established in the case of *Tarasoff v. Regents of the University of California*, there is a duty to protect the safety of an intended victim that supersedes the duty to maintain a patient's right to confidentiality. Anyone who has sexual relations with or engages in other risk behaviors with Scott is at risk for contracting HIV and syphilis, and protecting the health and welfare of those contact persons takes precedent over maintaining Scott's confidentiality. Additionally, if Scott knowingly exposes others to HIV, this may be considered a crime in some states, although individual state laws are variable regarding what qualifies as criminal activity in this regard. Health care practitioners should specifically learn the law of their state of practice regarding this issue. Due to variability in state laws regarding mandatory reporting of HIV status by patients to persons with occupational exposure to bodily fluids (e.g., dental hygienists, nurses, physicians, etc.), the use of universal contact precautions is always prudent.

The Ethics

The *Principle of Respect for Dignity* requires respect for Scott's emotions, his relationships, and his privacy. Scott is naturally upset about the possibility that his diagnosis of syphilis and HIV will be shared with his wife and he is worried about the effect this knowledge may have on their relationship. To protect Scott's relationship with his wife and his privacy, it appears that Scott's diagnosis should remain confidential and his wife should not be informed.

However, none of the principles of health care ethics is absolute and there are situations in which each of the principles may be overridden by more important considerations. In this case the *Principle of Respect for Dignity* and the medical confidentiality that it requires are outweighed by the need to protect others in society from grave and foreseeable harm. This conclusion can be justified by comparing the harms that would be done and the harms that would be prevented by violating confidentiality in this case. Scott will be harmed emotionally. He may feel angry and betrayed. His relationship with his wife may be ruined and may even end. On the other hand, if Scott's wife is informed, then she can get tested for syphilis and HIV, seek treatment if she has been infected, and if not infected, she can take steps to protect herself from becoming so. The harm prevented, then, is the harm of serious and life-threatening illness.

The same conclusion may be reached using the *Principles of Beneficence* and *Non-Maleficence*, considering the harms done and the harms prevented for Scott, alone. Informing Scott's wife of her exposure to syphilis and HIV means the loss of Scott's privacy and possibly his marriage, but it also prevents the emotional pain of knowing he has infected his wife with a deadly disease. Comparing these harms, it could be argued that notifying Scott's wife is in his best interests, as well as hers.

The same applies to notifying Scott's other sexual partners. Communicable diseases present a case in which individual privacy and medical confidentiality are overridden by public health concerns. This is the moral basis for the current reporting, counseling, and notification program carried out by local public health departments. Once Scott's diagnosis is reported, he will meet with a health investigator who will record the names of his contacts and notify them of their exposure to syphilis and HIV. Most states use "confidential" HIV testing, which means that contacts will not be told who has been diagnosed as HIV seropositive and has subsequently named them as contacts at risk for infection.

Scott should be informed that his diagnosis will be reported to the public health department and that he will be asked to meet with an investigator and give the names of his sexual contacts. Ideally, Scott would name his wife as a sexual contact and she would be notified along with his other contacts. However, it could be argued that it would be better for Scott's wife (in terms of time for tests and treatment or protection) and better for him (in terms of the future of their relationship) if Scott told her himself about his diagnosis and her risk. With this in mind, the medical resident could counsel Scott to inform his wife as soon as possible. Scott needs to be aware that, if they stay together for any length of time, his wife will likely come to know that he is very sick and she may guess the cause of his illness based on his symptoms and treatments.

If Scott refuses to tell his wife, the medical resident could intervene on the wife's behalf by informing the public health department that Scott is married, thereby guaranteeing she will be notified by the health investigator.

The Formulation

Now that the evidence-based medicine, legal precedent, and relevant ethical principles for this case have been reviewed, formulate a strategy to address the ethical conflicts in this case. If necessary, perform additional research into local and state laws and hospital regulations. Consider delving further into the background medical literature to assist with making sound therapeutic decisions. Devise a treatment approach that addresses the needs of the patient and his family, that is both ethically and medically sound, and that is culturally competent. Ensure that the strategy employs fair and appropriate utilization of medical resources, and that the approach is practical and feasible within the limits of the medical system at large. Work out a clear and professional way to communicate the proposal to the patient and his family. Attempt to foresee challenges that may arise in conveying or implementing the plan. Determine what follow-up will be necessary to ensure that the chosen strategy remains successful for the patient in the long-term. Reflect on how the knowledge and skills learned from this case can be used to improve the care of patients that may be encountered in future practice.

Afterthoughts

In this case a patient is diagnosed with two reportable infectious diseases and patient confidentiality must be broken to maintain the health and safety of those who have come into contact with him. How much does fear of "exposure" via broken confidentiality impact a patient's initiative to get tested for diseases such as HIV? Should HIV testing ever be anonymous or confidential? Is public health more important than an individual's health and rights?

Annotated References/Further Information

Code of Medical Ethics of the American Medical Association: Current Opinions with Annotations, 2006–2007 Edition. Council on Ethical and Judicial Affairs. Annotations prepared by the Southern Illinois University Schools of Medicine and Law.

Snyder L, JD, and Leffler C, JD. Ethics Manual, Fifth Edition. Ethics and Human Rights Committee, American College of Physicians. Annals of Internal Medicine. 142(7):560–582, 5 April 2005.

http://www.cdc.gov/hiv. Accessed October 15, 2007.

http://www.cdc.gov/hiv/topics/surveillance/basic.htm#hivest. Accessed October 15, 2007.

Glynn M and Rhodes P. Estimated HIV prevalence in the United States at the end of 2003. National HIV Prevention Conference; June 2005; Atlanta. Abstract 595.

Mocroft A et al. The incidence of AIDS-defining illnesses in 4,883 patients with human immuno-deficiency virus infection. Archives of Internal Medicine. 158:491–497; 9 March 1998.

CDC. Cases of HIV infection and AIDS in the United States, 2004. HIV/AIDS Surveillance Report 2005;16:16–45. Acquired from: *http://www.cdc.gov/hiv/topics/surveillance/resources/reports/2005report/default.htm.* Accessed October 15, 2007.

Tao G et al. Missed opportunities to assess sexually transmitted diseases in U.S. adults during routine medical checkups. American Journal of Preventive Medicine. 18(2):109–114. 2000.

http://www.cdc.gov/std/stats/syphilis.htm. Accessed October 15, 2007.

Ciesielski CA and Boghani S. HIV infection among men with infectious syphilis in Chicago, 1998–2000 [Abstract no. 12]. In: Program and abstracts of the 9th Conference on Retroviruses and Opportunistic Infections, Seattle, Washington, February 24–28, 2002. Acquired from: *http://www.retroconference.org/2002/Abstract/13221.htm.* Accessed October 15, 2007.

http://www.cdc.gov/hiv/topics/aa/resources/qa/downlow.htm. Accessed October 15, 2007.

http://www.epi.state.nc.us/epi/gcdc/manual/CDReportCard.pdf. Accessed October 15, 2007.

Case 13
When Care Becomes Futile

The Patient

Marie S. is a 67-year-old female with a longstanding history of poorly controlled hypertension who presented to the Emergency Department with her two adult daughters after developing an acute onset of right-sided weakness at home on the morning of admission. Her daughters first detected that her speech was abnormal, and that she was straining to speak even short sentences. They then noticed she had a mild right facial droop and flaccidity in the right upper and lower extremity. Paramedics were called and Marie was brought to the Emergency Department. History from Marie herself was limited due to her aphasia, although she did seem to understand and follow commands appropriately and she did, with some difficulty, state "head hurts" to the admitting physician. Physical exam was notable for an elevated blood pressure of 200/115, right-sided neglect, weakness and sensory loss, as well as the Broca's aphasia. Laboratory evaluation and EKG were within normal limits. A stat head CT showed ischemic change in the left temporal lobe. Due to the relatively large area of stroke on the CT, her elevated blood pressure, and duration of symptoms greater than three hours, it was felt that administering a thrombolytic agent was probably unsafe due to a high risk for hemorrhagic conversion of her stroke. Therefore, Marie was admitted to the neurology floor for further evaluation and medical management.

During frequent checks of her neurological examination over several subsequent hours, Marie was noted to have progressively worsening mental status. Her eyes were closed, she became completely aphasic, and she did not make any spontaneous movements or follow commands. A follow-up head CT showed a greatly increased region of ischemia punctated with several small areas of hemorrhage. Based on these findings, she was immediately transferred to the intensive care unit for closer monitoring.

From: *Evidence-Based Medical Ethics*
By: J.E. Snyder and C.C. Gauthier © Humana Press, Totowa, NJ

The Ethical Dilemma

The medical team met with Marie's daughters to explain the gravity of her medical condition and to discuss her code status. Marie had never completed legal documentation such as a Power of Attorney for Health Care or a Living Will. As a widow, her two daughters were her legal next of kin. When Marie's condition was explained to them, they thanked the doctors for the information and stated that "we know Jesus is watching her; He will protect her, and He will save her." The team expressed their sympathy for Marie's situation and offered to have a hospital chaplain become involved in the case, which the daughters accepted.

Over the next several days there was little change in Marie's status. When the team rounded on Marie each day the daughters were always by her bedside, often reading passages of the Bible to her. They reported that Marie would smile when they spoke to her and would grip their hands, and that "Jesus is healing her." However, when the team did a thorough neurological exam on Marie, these findings of clinical improvement were not objectively observed. The team was concerned that, despite optimal medical management, the likelihood of meaningful neurological recovery was slim, and there was still a high risk of mortality from the stroke itself or from a complication such as aspiration pneumonia. Additionally, the team believed that Marie would have a very poor chance of surviving a cardiopulmonary arrest, were it to happen, even if advanced cardiovascular life support (ACLS) protocols were implemented. Nevertheless, the daughters refused to consider these possibilities. They insisted that Marie remain a "full code" and that a temporary nasogastric feeding tube be placed for nutrition and medication administration "while she recovers."

Questions for thought and discussion: Do faith or other belief systems ever cloud a person's judgment, or is it the opposite – simply a matter of taking a different point of view in a dilemma? How can the medical team balance respect for other points of view (religious, cultural, etc.) and their own clinical beliefs for this patient's outcome?

Question for thought and discussion: How can 'futility' be objectively defined, if treatment goals and expectations may be different for a health care practitioner than for a patient and their family?

Marie began to have episodes of apnea and was subsequently intubated and placed on a mechanical ventilator. On the ventilator she was hemodynamically stable, but showed no objective signs of neurological improvement, and attempts to wean her from the ventilator were unsuccessful. When asked, Marie's daughters stated that they wanted her to undergo tracheostomy and percutaneous endoscopic gastrostomy (PEG) tube placement for her long-term support and care. Before these procedures could be performed Marie had a cardiac arrest and was not successfully resuscitated with ACLS.

Question for thought and discussion: Had Marie not had the cardiac arrest, should the medical team have endorsed tracheostomy and PEG tube placement for her?

The Medicine

According to statistics from the Centers for Disease Control and Prevention (CDC) there are about 700,000 strokes in the United States each year. Over 160,000 people die annually from stroke, making stroke the third leading cause of death nationally. For unclear reasons the highest stroke incidence and mortality is seen in the southeastern U.S., and increased mortality rates are observed in African-American and Native American patients.

Mortality is particularly high for patients with stroke that require mechanical ventilation, with a 30-day death rate estimated between 46 and 75 percent. Terminal extubation is commonly performed in patients with stroke, perhaps due to both the observed mortality rates and fundamental patient preferences. In a study of the latter, Patrick, et al. surveyed 341 persons about their wishes regarding life-sustaining interventions in the setting of a hypothetical severe stroke. Among those studied, 77 percent stated they would not want prolonged mechanical ventilation and 41 percent would not even want this in the short-term. Nearly two-thirds of respondents stated they would not want a feeding tube, hemodialysis, or cardiopulmonary resuscitation (CPR). In their study examining 105 patients that died of stroke in one neurointensive care unit, Mayer and Kossoff found 48 percent of patients had medical interventions withheld or withdrawn before dying, and 43 percent of those with cardiac (not brain) death had been terminally extubated. Of patients who were terminally extubated, the median survival was 7.5 hours and 69 percent died in the first 24 hours. There was a statistically significant higher frequency of decisions for terminal extubation among white and Hispanic patients than in African-American patients. In interviews with the health care agents making the decision to terminally extubate the patients, the vast majority were satisfied with their decision, felt that the patient had minimal suffering while dying, and would, in retrospect, probably make the same decision again.

If stroke-related death in the intensive care unit involves terminal extubation with as high a frequency as the Mayer and Kossoff study suggests, then accurately determining a futile prognosis is essential in these patients so that appropriate decisions are routinely made regarding withdrawal of support. In a review by Wijdicks and Rabinstein, several indicators of futility from various studies were identified (note that this is class III or IV evidence) and are summarized in Table 13.1.

The Law

In most states there is no legal protection for physicians who refuse to provide life-sustaining medical treatments they believe are futile when a patient or family member requests them. In the case, *In re the conservatorship of Helga M. Wanglie*, a district court judge appointed Mrs. Wanglie's husband as her conservator, even though he requested medical treatments the hospital staff and medical director believed were inappropriate. The judge ruled that Mr. Wanglie could best represent

Table 13.1 Clinical and radiological indicators, following massive stroke, of futility of care*

Type of stroke	Clinical profiles suggesting futility of care	Radiological profiles suggesting futility of care	Mixed clinical and radiological profiles suggesting futility of care
Subarachnoid hemorrhage from cerebral aneurysm	Coma persisting despite attempts to lower intracranial pressure	Massive intraventricular hemorrhage and hydrocephalus Delayed global edema on computed tomography (CT)	
Intracerebral hemorrhage (lobar)	Coma associated with extensor posturing and loss of pontomesencephalic reflexes		Coma associated with >6 mm septum pellucidum shift on CT
Intracerebral hemorrhage (ganglionic)			Coma with >60 cc hematoma volume and presence of hydrocephalus
Pontine hemorrhage	Coma associated with hyperthermia and tachycardia		Coma with thalamic extension of hemorrhage and acute hydrocephalus
Cerebellar hemorrhage	Absent corneal reflexes		Absent oculocephalic reflex and presence of hydrocephalus
Ischemic infarction (hemispheric)	Clinical deterioration including coma and absent pontomesencephalic reflexes	>4 mm shift of pineal gland on CT (that is performed within 48 hours)	
Ischemic infarction (cerebellar)	Persistent coma despite decompressive surgery		

Source adapted from: Wijdicks EF and Rabinstein AA. Absolutely no hope? Some ambiguity of futility of care in devastating acute stroke. Critical Care Medicine. 32(11):2332–42, 2004 November.
* Note that this is largely class III or IV evidence.

his wife's wishes. This case has sent the message that, when family members and other surrogate decision makers request life-sustaining medical treatments, they must be provided. Even without this message, physicians and hospitals are understandably reluctant to refuse to comply with requests for life-sustaining procedures. They are naturally concerned about the threat of a lawsuit and the charge of malpractice if they allow a patient to die when family members are asking for procedures that may preserve the patient's life.

Many hospitals have addressed this problem by formulating futile care or ineffective intervention policies. Most of these policies offer guidelines for physicians to use in determining when medical treatment is futile, information to include in discussions with family members, and hospital resources that may be helpful in resolving disagreements, such as the hospital ethics committee and risk management departments. Other options include transferring the patient to another physician or another hospital willing to accept the patient and to provide the disputed treatment.

There have been numerous attempts to define "futile care" (e.g., interventions that will not meet any reasonable medical goals for the patient, cannot reasonably be expected to bring about the patient's recovery or merely postpone the moment of death artificially). The American Medical Association Council on Ethical and Judicial Affairs recommends that denial of treatment should be based on "ethical principles and acceptable standards of care," rather than "the concept of futility, which cannot be meaningfully defined" (Code of Medical Ethics, 2.035). Without using the term "futile care," the Ethics Manual of the American College of Physicians states that physicians need not provide treatment when there is no evidence that it "will provide any benefit from any perspective." The more difficult situation, however, is when "the treatment will offer some small prospect of benefit at a great burden of suffering or financial cost, but the patient or family nevertheless desires it" (Snyder and Leffler, 2005).

In 1999 the Texas Advance Directives Act was revised so that physicians may refuse to provide life-sustaining medical treatments they believe are "inappropriate," even when a patient or surrogate decision maker requests them. Once a review committee has affirmed the physician's judgment that the requested treatment is inappropriate, the patient or surrogate is given 10 days to transfer the patient to another physician, another care setting in the same facility, or another facility. The physician and the hospital are "not obligated to provide life-sustaining treatment after the tenth day" once the patient or surrogate is notified in writing of the review committee's decision (Texas Statutes, Section 166.046).

The medical team has serious doubts about the benefits of a tracheostomy and PEG tube placement for Marie. They should inform Marie's daughters of these doubts and their reluctance to provide these procedures. However, before refusing to provide any life-sustaining medical treatment that is being requested by patients, their family members, or other legally-authorized surrogate decision makers, the medical team should become knowledgeable about the relevant policies in their hospital. They might also want to ask for an ethics committee consult or contact the risk management or legal department for advice on how to proceed.

The Ethics

It might appear that the *Principle of Respect for Autonomy* would require the medical team to provide any and all medical treatments requested by a patient or a surrogate decision maker, based on "substituted judgment," even if the team believes these treatments would be futile or inappropriate for the patient. However, this principle requires that capable patients must be permitted to accept or refuse *recommended* treatments, not that patients or their surrogates must receive any medical treatments they request or demand. If the medical team has not recommended a tracheostomy and PEG tube placement for Marie, this principle will not apply to the case.

Taken together, the *Principle of Beneficence* and the *Principle of Non-Maleficence* illustrate one of the conflicts created by requests such as those made by Marie's daughters. The American Medical Association Council on Ethical and Judicial Affairs writes that, "Physicians are not ethically obligated to deliver care that, in their best professional judgment, will not have a reasonable chance of benefiting their patients." Most practitioners think of benefits for their patients in terms of recovery, improvement, and relieving pain and suffering. They may consider interventions that merely prolong the dying process to offer no benefit to their patients. However, family members may consider continued life for their loved one to be a benefit, even when there is no consciousness and no chance of recovery.

The medical team determined that a tracheostomy and a PEG tube will not provide any benefit to Marie because they cannot bring about her recovery or change her medical condition. They may also believe that these interventions would cause unnecessary harm, considering the associated risks. On the other hand, Marie's daughters believe that continued life would be a benefit to Marie, while the healing they expect to occur can take place. Even if they are made aware of the risks of these interventions, they might consider them to be less important than the benefit of sustaining their mother's life. The conflict here concerns the identification of benefits and a comparison of benefits and risks of harm.

The *Principle of Respect for Dignity* points out another conflict that occurs when requests for inappropriate treatment are based on religious beliefs. Respecting the dignity of patients and their families requires that their religious and cultural backgrounds be respected. Marie's daughters want life-sustaining medical procedures provided to their mother because they believe she will recover based on their religious convictions. On the other hand, the dignity of medical practitioners is compromised when they are forced to provide medical treatment they have determined to be futile or inappropriate, based on their professional training and expertise.

One way to approach this conflict is to recognize that respect for the religious convictions of patients or their families does not require practitioners to accept them, to believe them, or to act according to them, in violation of their own professional judgment. What is required is that these beliefs are acknowledged and their contribution to decision making in the medical context is recognized. In this case, the medical team respected the religious beliefs of Marie's family by actively listening to the daughters and acknowledging the importance of their beliefs by offering the services of the hospital chaplain.

Based on the *Principle of Respect for Dignity*, the medical team should not make a judgment about the effects of the daughters' religious beliefs on their decision-making capacity. Moreover, these beliefs are not relevant to the question of what treatment the medical team should order for Marie. This decision should be based solely on the professional judgment of the team members.

According to the *Principle of Distributive Justice*, health care resources should be distributed in a fair way among members of society. This principle may apply when practitioners believe medical interventions will not improve the patient's condition or bring about recovery. Indefinite support on a ventilator and PEG tube, along with the total nursing care that is required, is a very expensive undertaking. Thus, practitioners may question whether such care for a patient who is neurologically devastated and will not recover is a good use of medical resources.

If Marie had not had a cardiac arrest the medical team should have continued their conversation with her daughters, emphasizing comfort care and pain relief for their mother, rather than medical interventions they would not recommend. Most importantly, they must be careful not to ask the daughters about procedures they believe are futile, ineffective, or inappropriate. In this case, someone asked the daughters about tracheostomy and a PEG tube and this may have given them the impression that these procedures were being offered or recommended for their mother. If inappropriate interventions are requested or demanded by family members, even when practitioners do not mention them, practitioners should be honest about their objections and their reasons.

Families and patients need the truth about the relevant diagnosis and prognosis, and guidance in thinking about reasonable medical goals in terms of benefits and risks of harm. Practitioners should also encourage family members to think about what their loved one would have wanted and about what would be in the patient's best interests regarding quality of life.

The medical team should also become familiar with the hospital's policy on futile care. This policy may include procedures for transferring the patient to another physician or another facility. The team may also consult the hospital ethics committee, the risk management department, or the hospital attorney for guidance. If hospital policy and/or these consultations advise that the requested treatment be provided, the medical team still does not need to "endorse" treatment to which they object. They can note in Marie's medical chart that they are providing the treatment under pressure and against their best medical judgment.

The Formulation

Now that the evidence-based medicine, legal precedent, and relevant ethical principles for this case have been reviewed, formulate a strategy to address the ethical conflicts in this case. If necessary, perform additional research into local and state laws and hospital regulations. Consider delving further into the background medical literature to assist with making sound therapeutic decisions. Devise a treatment approach that addresses the needs of the patient and her family, that is both ethically and medically sound, and that is culturally competent. Ensure that the strategy employs fair and appropriate utilization of medical resources, and that the approach is practical and feasible within the limits of the medical system at large. Work out a clear and professional way to communicate the proposal to the patient and her

family. Attempt to foresee challenges that may arise in conveying or implementing the plan. Determine what follow-up will be necessary to ensure that the chosen strategy remains successful for the patient in the long-term. Reflect on how the knowledge and skills learned from this case can be used to improve the care of patients that may be encountered in future practice.

Afterthoughts

In this case conflict occurred when a medical team felt that medical interventions for a patient were futile, but this contradicted the belief of the patient's family that "everything must be done" and that the power of faith was greater than the tools of medicine. How does one best balance decisions about *quantity* of life and *quality* of life? Is a practitioner ever truly able to separate their own personal belief system from the care of their patients? How necessary is this?

Annotated References/Further Information

CDC. Stroke Facts and Statistics. Acquired from: *http://www.cdc.gov/stroke/stroke_facts.htm.* Accessed October 15, 2007.

Holloway RG et al. Prognosis and decision making in severe stroke. JAMA. 294(6):725–33, 2005 Aug 10.

Wijdicks EF and Rabinstein AA. Absolutely no hope? Some ambiguity of futility of care in devastating acute stroke. Critical Care Medicine. 32(11):2332–42, 2004 Nov.

Mayer SA and Kossoff SB. Withdrawal of life support in the neurological intensive care unit. Neurology. 52(8):1602–1609. May 1999.

Patrick DL et al. Validation of preferences for life-sustaining treatment: implications for advance care planning. Annals of Internal Medicine. 127(7):509–517.

Code of Medical Ethics of the American Medical Association: Current Opinions with Annotations, 2006–2007 Edition. Council on Ethical and Judicial Affairs. Annotations prepared by the Southern Illinois University Schools of Medicine and Law. The subsection on Futile Care (2.035) was used in this case discussion.

Snyder L, JD, and Leffler C, JD. Ethics Manual, Fifth Edition. Ethics and Human Rights Committee, American College of Physicians. Annals of Internal Medicine. 142(7):560–582, 5 April 2005. The subsection on Making Decisions Near the End of Life contains an excellent discussion of this topic without using the term "futile care."

In re the conservatorship of Helga M. Wanglie, No. PX 91–283. District Probate Division, 4th Judicial District of Hennepin, State of Minnesota. Texas Statutes Title 2 Health, Chapter 166 Advance Directives, Section 166.046 Procedure if Not Effectuating a Directive or Treatment Decision.

Case 14
When Age Is a Factor in Health Care Decisions

The Patient

Robert W. is a 97-year-old male with a history of hypertension and mild congestive heart failure who presented to the Emergency Department complaining of the acute onset of dizziness on the morning of admission. Robert had been in his usual state of health until that morning when, shortly after arising from bed, his son noticed he was somewhat pale, weak-appearing, and generally "not himself." Paramedics were called and Robert was brought to the Emergency Department for further evaluation. By history Robert complained only of mild shortness of breath and generalized weakness. He denied any focal weakness or numbness, headache, or any other problems. At baseline Robert had, per his son's report, a relatively active lifestyle. He independently performed activities of daily living (ADLs) and was cognitively intact. Although he was widowed, he enjoyed spending time with his children and grandchildren, and frequently played cards with a close friend. His home medications included only an ACE inhibitor and a mild diuretic. He had a very remote history of 20-pack-years of cigarette smoking, but no alcohol or past illicit drug use.

On physical exam Robert was noted to be bradycardic with a heart rate of 45 beats per minute, but was normotensive and had an otherwise non-focal exam. Laboratory evaluation, head CT, and chest X-ray were all within normal limits. EKG showed sinus bradycardia in the 40s, as well as a left bundle branch block. Transthoracic echocardiography was notable only for diastolic dysfunction and mild, clinically insignificant valvular changes. He was admitted to a medical telemetry floor and a cardiology consult was obtained.

The Ethical Dilemma

After thorough evaluation the cardiologist diagnosed Robert with sick sinus syndrome, and recommended placement of a permanent dual-chamber pacemaker to maintain Robert's heart rate in the normal range. One of the medical residents questioned if a pacemaker should be offered to a 97-year-old.

From: *Evidence-Based Medical Ethics*
By: J.E. Snyder and C.C. Gauthier © Humana Press, Totowa, NJ

Questions for thought and discussion: How old is "too old" to offer medical interventions to a patient – is there ever an age cut off? Does the cost of the intervention ever matter?

The Medicine

Sick sinus syndrome (SSS), or sinoatrial node dysfunction, is a common cause of symptomatic bradycardia, with an estimated prevalence of one in 600 patients over age 65. Around 48 percent of all primary pacemaker implantations are performed for this condition. Sinoatrial node dysfunction can be a result of idiopathic fibrosis of the sinoatrial node, myocarditis, digitalis toxicity, the effects of other medications (e.g., β–adrenergic blockers, calcium channel blockers), cardiac surgery, electrolyte imbalance, hypothermia, hypothyroidism, or high vagal tone. Myocardial ischemia and infarction are also potential causes, although uncommon. Manifestations on the electrocardiogram include sinus bradycardia, sinus blocks and arrest, junctional or ventricular escape rhythms, atrial flutter or fibrillation, and the tachycardia-bradycardia syndrome. In patients with intrinsic, nonreversible causes of sinoatrial node dysfunction causing symptomatic bradycardia, permanent pacemaker implantation is generally considered the standard of care and is by and large successful in relieving the patient's symptoms. The choice of a dual-chamber pacemaker, rather than a single-chamber one, is supported in a 2004 Cochrane Database Systematic Review of the treatment of sick sinus syndrome and atrioventricular block.

The Law

There is no legal precedent for an age-based limitation on medical treatment in this country. However, financial considerations may be relevant to this case. In 2005 Rinfret, et al. estimated that the expense of the uncomplicated initial implantation of a rate-modulated dual chamber (DDDR) pacemaker – including the cost of the pacemaker and leads, the procedure, the hospitalization and physician fees – amounted to $11,203 in 2001 United States dollars. Adding the cumulative cost of follow-up medical care, medications and other subsequent expenses, the group estimated that a 74-year-old patient undergoing implantation of a DDDR pacemaker would incur $59,104 in lifetime costs, with a life expectancy of 6.49 years.

Given the patient's age in this case it may also be relevant to consider whether or not coverage would be provided by the federal Medicare program. According to the Centers for Medicare and Medicaid Services (CMS), cardiac pacemakers are covered as prosthetic devices, under certain conditions that are related to the patient's specific diagnosis. The CMS publishes a list of "Nationally Covered Indications" for use by Medicare contractors and providers for both the single-chamber and dual-chamber pacemaker. These indications may be found in the Medicare Coverage Database.

It appears that the placement of Robert's pacemaker would be covered under Medicare Part B, with the patient paying an annual deductible and 20 percent of the Medicare approved amount. Providers and suppliers may also charge more than the Medicare approved amount, adding to the patient's costs.

The Ethics

When the cardiologist recommended the placement of a pacemaker for Robert, a member of the medical team then asked if Robert was too old to receive a pacemaker. This is not a question of "futile care" or "ineffective interventions." The cardiologist has made this recommendation because Robert will benefit from the pacemaker as it will help maintain his heart rate within the normal range. Thus, the placement of the pacemaker is expected to meet reasonable medical goals for Robert.

In deciding about the best treatment to recommend to Robert, practitioners must consider the risks and benefits of the pacemaker. Using the *Principle of Beneficence*, the pacemaker is expected to prevent the harm of death from a cardiac arrest, to remove the harm caused by Robert's sick sinus syndrome, and to promote the good of proper cardiac functioning. Practitioners would also consider the side effects of the placement, operation, and maintenance of the pacemaker to avoid doing harm, using the *Principle of Non-Maleficence*.

The cardiologist has presumably made these calculations and concluded that the potential benefits of the pacemaker outweigh the possible side effects for Robert, so the pacemaker would be in Robert's best interests. It seems that the medical team is not questioning the cardiologist's comparison of risks and benefits, but there is a concern about Robert's age.

It might appear, at first glance, that age *is* a relevant consideration for such an invasive medical intervention. The usual image of a 97-year-old man may be to some that of a frail, elderly patient suffering from a variety of physical ailments, losing his cognitive capacities, and dependent on others for basic personal care. Placing a pacemaker may be questionable for the "typical" 97-year-old patient, but Robert does not have other serious health problems. He is active and independent, able to care for himself, and does not have any cognitive deficits. If the sick sinus syndrome is Robert's only serious health problem at this time, and the pacemaker will provide clear benefits in treating this problem, it will be difficult to argue against placing it. Comparing Robert's case with that of the "typical" patient in his age group indicates that the patient's quality of life and general health will be much more relevant when deciding which medical interventions to offer than will age alone.

Robert is capable of considering the risks and benefits of the pacemaker placement and follow-up medical care, including medications, and he is capable of deciding whether to accept or refuse this intervention. Based on the *Principle of Respect for Autonomy*, unless the medical team has a reason (other than Robert's age) to question the cardiologist's recommendation, the choice to undergo the pacemaker placement should be left up to Robert.

According to the requirements of voluntary informed consent Robert will need to be given full information about the benefits and risks of the pacemaker placement and follow-up care, and any reasonable alternatives, as well as the expected outcome of no treatment. This information should be provided in language Robert can understand. He should not be pressured or subjected to undue influence to accept or refuse the pacemaker or any treatment alternatives.

Under the *Principle of Distributive Justice*, health care resources should be distributed in a fair way among the members of society. It may be argued that offering a pacemaker to Robert would be a poor use of costly resources because of his relatively short life expectancy at age 97. On the other hand, to deny him the pacemaker would surely shorten his life and reduce his quality of life, which is now unusually good for a man of his age. It also seems inappropriate for individual practitioners to deny their patients clearly beneficial medical interventions based on either age or financial considerations. As the Ethics Manual of the American College of Physicians points out, "Resource allocation decisions are most appropriately made at the policy level rather than entirely in the context of an individual patient-physician encounter." Moreover, our policy makers have determined that age will not be a factor in the coverage of cardiac pacemakers by public funds, under the federal Medicare program. The conditions and limitations for cardiac pacemakers listed in the Medicare Coverage Database concern the results of diagnostic tests and the patient's specific diagnosis.

A final justice question is raised by the fact that patients, or their insurance companies, are responsible for 20 percent of the pacemaker's cost or more if providers or suppliers charge more than the Medicare-approved amount for this intervention and subsequent follow-up care. If Robert is able to pay these costs, but other patients with equal need, in good general health, and with a good quality of life are not, is it fair for Robert to get a pacemaker while the other patients may not?

The Formulation

Now that the evidence-based medicine, legal precedent, and relevant ethical principles for this case have been reviewed, formulate a strategy to address the ethical conflicts in this case. If necessary, perform additional research into local and state laws and hospital regulations. Consider delving further into the background medical literature to assist with making sound therapeutic decisions. Devise a treatment approach that addresses the needs of the patient and his family, that is both ethically and medically sound, and that is culturally competent. Ensure that the strategy employs fair and appropriate utilization of medical resources, and that the approach is practical and feasible within the limits of the medical system at large. Work out a clear and professional way to communicate the proposal to the patient and his

family. Attempt to foresee challenges that may arise in conveying or implementing the plan. Determine what follow-up will be necessary to ensure that the chosen strategy remains successful for the patient in the long-term. Reflect on how the knowledge and skills learned from this case can be used to improve the care of patients that may be encountered in future practice.

Afterthoughts

In this case a question arose about a patient's age influencing decisions about their care. One often observes in medical practice that a patient's age does not necessarily correlate with their quality of life and the morbidity of their underlying disease processes. For example, a very ill 50-year-old may have a worse quality of life than a vibrant 80-year-old. These two patients may also have different treatment expectations from their health care providers. Despite this, health care providers may make inappropriate assumptions about an older patient's expectations for health care, such as code status or the desire to undergo certain treatments and procedures. Moreover, many research studies place age limitations on participants, so the benefit of offering certain interventions for older patients is unclear. As the United States population ages, more of these health care concerns will likely come to the forefront of ethical debates.

If age is not a determinant for qualifying for certain medical treatments, should quality of life be? Should quality of life be determined by an objective scale, calculated by a health care practitioner, or should it be based on a patient's perception of it alone? How might either of these options create ethical conflicts?

Annotated References/Further Information

Code of Medical Ethics of the American Medical Association: Current Opinions with Annotations, 2006–2007 Edition. Council on Ethical and Judicial Affairs. Annotations prepared by the Southern Illinois University Schools of Medicine and Law.

Snyder L, JD, and Leffler C, JD. Ethics Manual, Fifth Edition. Ethics and Human Rights Committee, American College of Physicians. Annals of Internal Medicine. 142(7):560–582, 5 April 2005. The subsection on Resource Allocation includes two principles that should guide physicians as they face the ethical problems raised by the just allocation of resources and changing reimbursement methods. The quote included in this case is the second of those principles.

Mangrum JM and DiMarco JP. The evaluation and management of bradycardia. New England Journal of Medicine. 342(10):703–9, 2000 Mar 9.

Bernstein AD and Parsonnet V. Survey of cardiac pacing in the United States in 1989. The American Journal of Cardiology. 1992;69:331–8.

Dretzke J et al. Dual chamber versus single chamber ventricular pacemakers for sick sinus syndrome and atrioventricular block. Cochrane Database of Systematic Reviews. 2004.

Da Costa D et al. Bradycardias and atrioventricular conduction block. British Medical Journal. 324(7336):535–8, 2002 Mar 2.

Department of Health and Human Services, Centers for Medicare and Medicaid Services. Publication 100–03, Medicare National Coverage Determinations. Section 20.8 Cardiac Pacemakers. *www.cms.hhs.gov/Transmittals/Downloads/R16NCD.pdf.* Accessed October 15, 2007.

Rinfret, S. et al. Cost-effectiveness of dual-chamber pacing compared with ventricular pacing for sinus node dysfunction. Circulation. 2005;111:165–172.

Case 15
When a Patient Is Unidentifiable

The Patient

John Doe is an approximately 50-year-old male with an unknown past medical history who was brought to the Emergency Department after being found unresponsive in an alleyway by passersby. The patient was placed on a backboard, intubated, and had a hard cervical collar placed prior to E.D. arrival due to outward signs of significant physical injury. He was disheveled, had blood- and dirt-stained clothes, and had no identifying information on his person. A trauma survey revealed fractures to the pelvis and bilateral femurs, splenic and liver lacerations, a left hemothorax, and a large left-sided subdural hematoma – findings suggestive that he was likely a pedestrian victim of a (hit-and-run) motor vehicle collision. The patient went emergently to the operating room for procedures related to the above injuries, including drainage of the subdural hematoma.

After surgical stabilization, the patient was transferred to the surgical-trauma intensive care unit for further management. Police investigating the case were unable to identify the patient as he did not match the reports of any known missing persons. Due to his general appearance, they suggested that the patient may be homeless and that his identification may be difficult or impossible. Fingerprinting was done to potentially match existing criminal records in a national database, but this endeavor was not successful.

The Ethical Dilemma

Although John Doe was not consented for his initial surgical interventions, these were emergency, life-saving procedures done on an unconscious patient, so informed consent was waived. However, once the patient was stabilized surgically and was not in imminent danger of death, he required various non-emergent procedures, such as obtaining central venous access with a triple lumen catheter. He remained unable to consent for these procedures due to poor neurological status and inability to be weaned from the mechanical ventilator, and no one had yet come

From: *Evidence-Based Medical Ethics*
By: J.E. Snyder and C.C. Gauthier © Humana Press, Totowa, NJ

forward to identify him. The surgical trauma team obtained an ethics consult to determine what procedures may be done on the patient without his consent, whether to institute nutrition through a percutaneous endoscopic gastrostomy (PEG) tube, and how to best make other care decisions for him in the future.

Question for thought and discussion: Under what circumstances is informed consent not necessary in the care of a patient?

Questions for thought and discussion: Who becomes the health care agent for a patient without an identity? How does a designated health care agent make decisions in a patient's "best interest" when the agent does not know anything about the patient or their belief system?

Question for thought and discussion: Can supportive care be withdrawn from a patient in this situation if their likelihood of recovery is poor, or are they potentially destined for a lifetime of ventilatory and nutritional support?

The Medicine

In patients with insults to the brain or spinal cord, intubation and mechanical ventilation may be necessary for neurological reasons, and not for underlying pulmonary disease. Higher rates of reintubation, prolonged courses of mechanical ventilation, and greater need for tracheostomy may be observed in these patients. Additionally, the usual methods of liberating the patient from the ventilator by checking "weaning parameters" using respiratory measurements alone may need to be supplemented by assessments of neurological function. In one study of 100 neurosurgical patients by Namen, et al., a Glasgow Coma Scale (GCS) of less than eight was associated with a 33 percent success rate of extubation. However, a GCS score of greater than eight had a 75 percent success rate. In patients requiring long-term mechanical ventilation, tracheostomy and placement of a percutaneous endoscopic gastrostomy (PEG) tube for nutrition may be necessary.

The Law

The U.S. Court of Appeals' decision in *Canterbury v. Spence* set a precedent for the legally recognized exceptions to informed consent. The first exception occurs "when the patient is unconscious or otherwise incapable of consenting, and harm from a failure to treat is imminent and outweighs any harm threatened by the proposed treatment." It is this exception to informed consent that is applicable in the case of John Doe. The harm that would result from a failure to treat the patient doesn't have to be death, but it must outweigh the harm that might be caused by the treatment. In this case the central line was required to provide needed medications to the patient and to perform blood tests necessary for monitoring his condition. In order to determine whether or not they should place a central line, without consent, the surgical team needs to compare the harm that might be caused by this procedure with the harm that would come to the patient without the procedure.

Whatever decision is made about the central line, the surgical team should contact the hospital legal department immediately so that a formal application for a legal guardian for this patient may be initiated. In most states a judge will determine the need for a guardian and the department of social services may also be involved. In cases such as this the patient is likely to become a "ward of the state." The guardian will become John Doe's surrogate decision maker and will be expected to make medical decisions that they feel will be in the patient's best interests. The decision about PEG tube placement, for example, should probably be postponed until a guardian has been appointed and has met with the medical team about the patient's medical situation.

The Ethics

Based on the *Principle of Beneficence*, the patient's injuries have been treated to repair the damage caused by the collision, alleviate his pain, and prevent immediate death. It was also important to sustain his life on the ventilator, while the extent of his injuries and his chances for recovery could be determined. Based on the *Principle of Non-Maleficence* the risks of these interventions causing harm were also considered and found to be outweighed by the immediate benefits. The medical team is now considering further medical procedures that would be necessary for the patient's on-going treatment and to sustain his life over a longer period of time.

Because of John Doe's poor neurological status, the *Principle of Respect for Autonomy* will not apply to this case. The patient is not able to make his own decisions and, because he cannot be identified, it is not possible to contact his family. Without a family member to act as surrogate decision maker, there is no way to consider the patient's wishes regarding medical treatment. Thus, the standard of "substituted judgment" cannot be used to make these decisions.

Decisions about treatment for this patient will need to be made using the "best interests standard." A guardian appointed by the court to make medical decisions for a complete stranger is at a disadvantage, compared with family members who have known the patient for years. However, the basic elements of this standard remain the same regardless of who is making these decisions and their relationship with the patient.

The "best interests standard" is based on the *Principles of Beneficence* and *Non-Maleficence*. What is in the patient's best interests will be based on the benefits and risks of the treatment under consideration; in particular, the likelihood that the patient will recover or improve, with treatment, compared with the possible harmful side effects. Other considerations will include the pain and suffering the patient may experience and the quality of life the patient may attain, in terms of consciousness, mobility, and dependence.

The medical team should meet with John Doe's guardian, once a guardian has been appointed, and explain his medical situation and any recommended procedures and treatments they believe would be in the patient's long-term best interests, now that he is out of immediate danger. The medical team should determine whether continued ventilator support and PEG tube placement would be effective in meeting any reasonable goals for this patient or would merely artificially postpone the moment of death. If they decide that these interventions would be ineffective or inappropriate, they should explain their reasoning and their reluctance to provide what they believe to be futile treatment.

The *Principle of Respect for Dignity* is particularly relevant in cases like this, where the patient cannot speak for himself and no one is available to speak for him. Because of his vulnerability, John Doe's privacy and bodily integrity should be protected as far as possible, while providing appropriate and beneficial medical care. This means, in part, that invasive life-sustaining medical treatment should not be initiated or continued simply because the patient's own wishes cannot be known. If the guardian agrees with the recommendations of the medical team, for example, not to place a PEG tube or to withdraw the ventilator, the patient need not be continued on life support.

On the other hand, the *Principle of Distributive Justice* should not be used to withdraw or withhold medical treatment just because the patient cannot be identified. As argued above, when no one knows what the patient would want, the standard for medical decision making should always be the "best interests standard."

The Formulation

Now that the evidence-based medicine, legal precedent, and relevant ethical principles for this case have been reviewed, formulate a strategy to address the ethical conflicts in this case. If necessary, perform additional research into local and state laws and hospital regulations. Consider delving further into the background medical literature to assist with making sound therapeutic decisions. Devise a treatment approach that addresses the needs of the patient, that is both ethically and medically sound, and that is culturally competent. Ensure that the strategy employs fair and appropriate utilization of medical resources, and that the approach is practical and feasible within the limits of the medical system at large. Work out a clear and professional way to communicate the proposal to the patient's legal guardian. Attempt to foresee challenges that may arise in conveying or implementing the plan. Determine what follow-up will be necessary to ensure that the chosen strategy remains successful for the patient in the long-term. Reflect on how the knowledge and skills learned from this case can be used to improve the care of patients that may be encountered in future practice.

Afterthoughts

In this case a medical team was challenged in caring for a patient with an unknown identity and no one to represent him when he himself did not have the capacity to make health care decisions. When patients become "wards of the state," how do their representatives ensure that good decisions are made on their behalf? To what extent is the representative going to rely on their own belief system to make important decisions? Is this ethically sound?

Annotated References/Further Information

Duguet A et al. Predicting the recovery of ventilatory activity in central respiratory paralysis. Neurology. 67(2):288–92, 2006 Jul 25.

Namen AM et al. Predictors of successful extubation in neurosurgical patients. American Journal of Respiratory and Critical Care Medicine. 163(3 Pt 1):658–64, 2001 Mar.

Manthous CA et al. Liberation from mechanical ventilation: A decade of progress. Chest 1998; 114:886–901.

Canterbury v. Spence (1972). U.S. Court of Appeals, District of Columbia Circuit. 464 Federal Reporter, 2nd series, 772.

Code of Medical Ethics of the American Medical Association: Current Opinions with Annotations, 2006–2007 Edition. Council on Ethical and Judicial Affairs. Annotations prepared by the Southern Illinois University Schools of Medicine and Law.

The subsection on Informed Consent (8.08) includes a similar exception to that found in *Canterbury v. Spence*.

Snyder L, JD, and Leffler C, JD. Ethics Manual, Fifth Edition. Ethics and Human Rights Committee, American College of Physicians. Annals of Internal Medicine. 142(7):560–582, 5 April 2005. In the subsection on Informed Consent the Ethics Manual recognizes the need for court involvement "to establish guardianship for an unbefriended, incompetent patient."

Case 16
When Next-of-Kin Disagree

The Patient

Ann C. is a 69-year-old African-American female with a past medical history of poorly controlled hypertension, diabetes, atrial fibrillation, and dyslipidemia who presented to the Emergency Department after being found unresponsive at home by her neighbor. No history could be obtained from Ann on presentation as she was only uttering incomprehensible sounds. Ann's outpatient chart indicated that warfarin was one of her home medications.

On physical exam she was noted to be hypertensive (210/99), bradycardic (heart rate of 50 beats per minute), and to be flaccid in all four extremities. There were coarse, wet breath sounds at both lung bases, suggestive of an aspiration event. Additionally she did not open her eyes to verbal command, but did withdraw from noxious stimuli (i.e., Glasgow Coma Scale of 7 out of 15). Laboratory evaluation was notable for a slight leukocytosis, mild acute renal failure consistent with volume depletion, and an International Normalized Ratio (INR) of 3.4. Electrocardiogram was remarkable only for left ventricular hypertrophy and nonspecific T-wave flattening. Non-contrast CT scan of the brain demonstrated that the patient had sustained a devastatingly large intracerebral hemorrhage that was supratentorial and extended into the lateral ventricles. There was mass effect from edema, but no evidence of herniation.

Ann's anticoagulation was quickly reversed with fresh frozen plasma (FFP) and a nicardipine drip was started for her hypertension. A neurosurgeon was consulted, who determined that the hemorrhage was not amenable to surgical intervention, and continued medical management in the intensive care unit was recommended. Based on available scoring methods for predicting prognosis, the neurosurgeon stated that Ann's risk of inhospital mortality from the event approached 100 percent.

From: *Evidence-Based Medical Ethics*
By: J.E. Snyder and C.C. Gauthier © Humana Press, Totowa, NJ

The Ethical Dilemma

Ann had lived at home for the 10 years prior to admission with her identical twin sister, Isabel. Both Ann and Isabel were well-known and liked by the medical team, since they always accompanied each other to well visits in the outpatient medical clinic and were always very pleasant. At the time of Ann's admission, Isabel stated that Ann would never want "to be kept alive artificially using life support" per her own stated wishes in the past. However, Ann had never filled out a Living Will or designated anyone as her health care agent on a Power of Attorney for Health Care. To complicate matters, Ann had three adult children who lived out of state. They only spoke to Ann, on average, once a year by phone around the holidays, and none had ever spoken with their mother about end-of-life issues. When they were notified of Ann's condition the two oldest children stated that they wished that Ann be kept alive on a mechanical ventilator, contrary to what Isabel stated about Ann's wishes. They affirmed that they wanted "everything done" to try and save her life. When the youngest child was contacted, she stated that she did not know her mother's wishes on this topic, but affirmed "I know that no one would want to be kept alive like that."

> **Questions for thought and discussion:** Who gets to make the decision for life support in this case? How does the law's definition of next of kin differ from what ethically may be the "best" choice?

> **Question for thought and discussion:** As today's definition of "family" has evolved from the definition of family from 50 years ago, do you see laws regarding next-of-kin evolving as well?

> **Questions for thought and discussion:** What is in Ann's best interest with regard to artificial life support or being allowed to pass away without such medical intervention? How does the practitioner's own belief system about the dying process play a role in how they can act in Ann's best interest?

Ann's condition continued to deteriorate. She was placed on a mechanical ventilator and remained a "full code" as per the wishes of her oldest children, since they were determined by the medical team to legally be Ann's next of kin. Repeat CT of the brain several hours later showed expansion of the hematoma and new evidence of herniation. Despite optimal support Ann had an asystolic cardiac arrest shortly afterwards and attempts at resuscitation were unsuccessful.

> **Question for thought and discussion:** Is it possible that race, ethnicity, or other cultural factors were influences in how Ann's condition was viewed by her family and how end-of-life decisions were made?

The Medicine

Atrial fibrillation is a common problem seen in clinical practice, and long-term anti-thrombotic therapy is often employed to reduce the risk of atrial fibrillation-associated stroke. However, nearly 12 percent of the 60,000 intracerebral hemorrhages occurring annually in the United States are associated with the use of these agents. Of these adverse events about 57 percent occur in aspirin users and 43 percent are in patients on warfarin. The risk of intracerebral hemorrhage is highest in patients with advanced age, history of prior stroke, hypertension, brain neoplasm, vasculitis, aneurysms, and other vascular malformations.

Overall, stroke is the third leading cause of death in the United States. The Centers for Disease Control and Prevention (CDC) report that there are approximately 700,000 strokes each year in the United States, with more than 160,000 annual deaths due to stroke. Nationally, the highest stroke incidence and mortality is seen in the southeastern United States, although the reasons for this are unclear. Increased mortality rates are also observed in African-American and Native American patients.

The 30-day mortality rate for patients with stroke that require mechanical ventilation is estimated to be between 46 and 75 percent. In patients with lobar intracerebral hemorrhage, factors such as presence of coma, extensor posturing, absent pontomesencephalic reflexes, and septum pellucidum shift >6mm on computed tomography are all highly predictive of futility of care after devastating stroke. The APACHE (acute physiology and chronic health evaluation) scoring system, is an additional method of predicting mortality in critically ill patients, such as those with multi-system organ failure.

The Law

The hierarchy of possible surrogate decision makers for a patient who is not capable of making medical decisions is established by state law. In 32 states, when no health care agent has been appointed by the patient, the surrogate will be: the patient's legal guardian, spouse, adult child, parent, or adult sibling, in that order (see Table 16.1). In 12 states no priority is specified.

In most states, then, Ann's adult children will have the legal authority to make medical decisions for her. When there is disagreement, as in this case, the decision to which the majority of children agree will be the controlling decision. However, this does not necessarily make Ann's two oldest daughters the best surrogate decision makers, particularly if their lack of recent interaction with and knowledge about their mother's wishes are considered.

If this case took place in one of the 12 states where no priority is specified for surrogate decision makers, the medical team could argue that Ann's sister, Isabel,

Table 16.1 The hierarchy of possible surrogate decision makers for a patient*

The health care agent listed on a Power of Attorney for Health Care
↓
The patient's legal guardian
↓
The patient's legal spouse
↓
The adult child(ren) of the patient
↓
The parent(s) of the patient
↓
The patient's adult sibling(s)
↓
Other persons (e.g., relatives, friends) well-acquainted with the patient, who have had reasonable contact with them and demonstrate a good understanding of their beliefs and wishes

*In 32 states. Please see individual state laws for proper determination of the surrogate decision maker as it varies from state to state, and this is an area of constantly evolving legislation.

should make these decisions for her, citing that the sisters have lived together for the past 10 years, so Isabel knows more about what Ann would want in terms of life-sustaining medical treatment. Therefore, Isabel could employ "substituted judgment," making a decision that would be similar to what Ann would decide for herself, if she were capable.

There is some evidence that laws regarding surrogate medical decision making are changing to reflect an expanded conception of "family." For example, in Arizona, "domestic partner" is on the hierarchy of possible surrogates, after adult child and parent. In New Mexico "significant other" is listed after spouse, but before adult child and parent in the hierarchy of surrogate decision makers. In Colorado and Hawaii medical decisions for those patients who are not capable of making them and without a health care agent are made by a consensus of "interested parties." As noted above, in 12 states no priority is specified and this may allow practitioners to make a case for someone who has not traditionally been considered a "family member," but who would know best what the patient would want at the end of life.

The Ethics

This is a case where the law and ethics disagree. Ann has lived with her twin sister for 10 years. They have been very involved in each other's health care, routinely accompanying each other to clinic visits. Isabel has the intimacy, knowledge, and empathy to be the best surrogate decision maker for her sister, Ann. Isabel has shared knowledge about what Ann would want with the medical team. If she were the surrogate she could make end-of-life decisions for her sister based on "substituted judgment," since she knows what her sister would want. In this way, the medical team would be respecting Ann's autonomy.

One disturbing element of this case is that Ann's daughters are making life and death decisions about someone they haven't seen in years and speak to by phone about once a year. They have no way of knowing what their mother would want in terms of end-of-life care. The daughters may be asking to have "everything done" out of guilt, perhaps because they had not kept in touch with their mother and do not visit her regularly. They may be thinking they can make up for years of neglect by keeping their mother alive.

Even worse, they are making these decisions from far away, without seeing their mother in her present condition. Certainly, before the daughters are permitted to make any medical decisions for their mother, they must come to the hospital to see her as she is now and become fully informed about her devastating stroke and prospects for the future. The daughters must be educated about this diagnosis and its prognosis. They need to hear the medical team's recommendations and the reasons behind them. They may even need counseling to accept their mother's dire prognosis. None of this can be done over the phone.

The medical team will recommend to Ann's daughters what they believe to be in Ann's best interests, based on the *Principles of Beneficence* and *Non-Maleficence*. There is no chance for recovery or long-term survival for Ann. The medical team may explain that the ventilator and resuscitation will not promote any long-term good for Ann and will cause the harm of bodily invasion, while simply prolonging the dying process. They could point out that a natural death, without aggressive and invasive medical procedures, would actually be in Ann's best interests.

The *Principle of Respect for Autonomy* will not apply in this case, unless the daughters are willing to consider what their Aunt Isabel said about their mother's own wishes regarding life support. According to the *Principle of Respect for Dignity*, the patient's relationships, reasonable goals, and bodily integrity should be recognized and taken into consideration as decisions are being made. It appears that Ann's relationship with her twin sister was important to her and one of her medical goals was to die naturally and without life-sustaining interventions. It is also true that these interventions will be invasive to Ann's body.

The *Principle of Respect for Dignity* also requires respect for the family's social, religious, and cultural background. What is both interesting and difficult about this case, is that the family is divided on what that background means in terms of medical decisions for Ann. Isabel and the youngest daughter seem comfortable with allowing Ann to die a natural death. Isabel is basing this on what she believes Ann would want. The youngest daughter is basing this on what she believes most people would want. On the other hand, the two oldest daughters want "everything done," and they offer no reasons for this, other than it is what they want.

Members of the medical team should try to set aside their own religious and cultural beliefs concerning life and death and practice medicine according to scientific evidence and standards of care established by medical science. According to the *Principle of Respect for Dignity*, the medical team should also resist demands by Ann's daughters for what the team believes are futile or inappropriate interventions.

If Ann's daughters continue to make these demands, the medical team could ask them why they want "everything done," and what treatments and procedures they imagine this to include. They could ask if they think being kept alive is what their mother would want and, if so, what reason they have to believe this. These kinds of questions may even help the daughters understand their own feelings about their mother's imminent death.

With regard to the influence of race and ethnicity on end-of-life decisions, it is important to define these two terms. *Race* is an arbitrary classification based on physical (phenotypic) characteristics and comprises a group of persons related by common descent or heredity. *Ethnicity* is an arbitrary classification based on cultural, religious, or linguistic traditions, in addition to ethnic traits, background, allegiance, or association. The U.S. Food and Drug Administration (FDA) has adopted federal standards for classifying race for practical purposes, and includes the following seven groups: American Indian or Alaska Native, Asian, Black or African-American, Native Hawaiian or Other Pacific Islander, White, Some Other Race, and Two or More Races. Additionally, the following groups of ethnicity are classified: Hispanic or Latino; Not Hispanic or Latino. This federal classification system stipulates that patients *self-designate* with a particular racial or ethnic group. The term *culture* best embodies the belief systems, attitudes, and behaviors that are characteristic of social, ethnic, age, or other groups. Race and ethnicity are often useful markers for identifying the cultural background and belief system of a patient.

Ann and her family self-identified as African-American. Studies have shown that African-American families expect the death of a functionally impaired family member less often than white non-Hispanic and white Hispanic families. This may be interpreted as inappropriate optimism by a white health care practitioner, simply due to the different belief systems amongst the two groups. Additionally, patients and families belonging to some minority racial and ethnic groups may choose more aggressive medical interventions, such as the use of life-sustaining machines at the end of life, than white patients and families. One study by Blackhall, et al. suggested that although African-Americans felt that withholding or withdrawing life support from a given patient was acceptable, they were more likely than members of the other ethnic groups studied to personally want to be kept alive on life support. Reasons for this difference, based on interviews with the African-American participants, included a general distrust towards the health care system and a fear that health care choices were based on one's ability to pay. Many other similar papers have shown that black patients are less likely to limit the use of life-extending treatments than their white counterparts. It has been proposed that the past disenfranchisement of blacks, in history and through medical incidents such as the Tuskegee experiments, has created a perception among black patients that limiting them from receiving life-sustaining care is equivalent to an injustice. Health care providers must be aware of racial, ethnic, and cultural differences when assisting patients and families through the medical decision-making process. This is especially true in the setting of end-of-life care.

The Formulation

Now that the evidence-based medicine, legal precedent, and relevant ethical principles for this case have been reviewed, formulate a strategy to address the ethical conflicts in this case. If necessary, perform additional research into local and state laws and hospital regulations. Consider delving further into the background medical literature to assist with making sound therapeutic decisions. Devise a treatment approach that addresses the needs of the patient and her family, that is both ethically and medically sound, and that is culturally competent. Ensure that the strategy employs fair and appropriate utilization of medical resources, and that the approach is practical and feasible within the limits of the medical system at large. Work out a clear and professional way to communicate the proposal to the patient and her family. Attempt to foresee challenges that may arise in conveying or implementing the plan. Determine what follow-up will be necessary to ensure that the chosen strategy remains successful for the patient in the long-term. Reflect on how the knowledge and skills learned from this case can be used to improve the care of patients that may be encountered in future practice.

Afterthoughts

In this case conflict arose because the person who probably knew Ann the best, and the longest, did not have legal rights to make decisions on Ann's behalf. Additionally, various family members differed how best to manage Ann's case. The case may also have been further complicated by racial, ethnic, and cultural differences between the patient's family and the medical team.

Would this case have been different if Ann had been engaged to be married? What rights would her fiancé have? What if Ann had a same-sex partner for many years? What if she had married her same-sex partner, but lived in a state where that marriage was not legally recognized? As a rule, how can a health care practitioner best follow the laws regarding next-of-kin as well as act in their patient's best interest in such situations?

Annotated References/Further Information

Code of Medical Ethics of the American Medical Association: Current Opinions with Annotations, 2006–2007 Edition. Council on Ethical and Judicial Affairs. Annotations prepared by the Southern Illinois University Schools of Medicine and Law.

Snyder L, JD, and Leffler C, JD. Ethics Manual, Fifth Edition. Ethics and Human Rights Committee, American College of Physicians. Annals of Internal Medicine. 142(7):560–582, 5 April 2005.

CDC. Stroke Facts and Statistics. Acquired from: *http://www.cdc.gov/stroke/stroke_facts.htm*. Accessed October 15, 2007.

Holloway RG et al. Prognosis and decision making in severe stroke. JAMA. 294(6):725–33, 2005 Aug 10.

Wijdicks EF and Rabinstein AA. Absolutely no hope? Some ambiguity of futility of care in devastating acute stroke. Critical Care Medicine. 32(11):2332–42, 2004 Nov.

Knaus WA et al. The APACHE III prognostic system: Risk prediction of hospital mortality for critically ill hospitalized adults. Chest. 1991;100:1619–1636.

Hart RG et al. Avoiding central nervous system bleeding during antithrombotic therapy. Recent Data and Ideas. Stroke. 2005;35:1588–1593.

Gorelick PB and Weisman SM. Risk of hemorrhagic stroke with aspirin use. An update. Stroke. 2005;36:1801–1807.

Ruiz-Sandoval JL et al. Grading scale for prediction of outcome in primary intracerebral hemorrhages. Stroke. 2007;38:1641–1644.

Ferdinand KC and Armani, A. Cardiovascular risk and disease in ethnic and racial groups. In Gotto/Toth (Eds.) Comprehensive Management of High-Risk.

Cardiovascular Patients. 2006. New York: Informa Health.

Williams BA et al. Functional impairment, race, and family expectations of death. Journal of the American Geriatrics Society. 54:1682–1687, 2006.

Blackhall LJ et al. Ethnicity and attitudes towards life sustaining technology. Social Science & Medicine. 1999;48:1779–1789.

Degenholtz HB et al. Race and the intensive care unit: Disparities and preferences for end-of-life care. Critical Care Medicine. 2003 Vol. 31, No. 5 (Suppl.): S373–S378.

"Culture." (n.d.). Dictionary.com Unabridged (v 1.1). Retrieved July 29, 2007, from Dictionary. com website: *http://dictionary.reference.com/browse/culture*.

The following website features a United States map illustrating the hierarchy for surrogate decision making in the different states:

www.time.coma/time/covers/1101050404/schiavo_webguide.html. Accessed October 15, 2007.

Case 17
When a Mistake Has Been Made

The Patient

Joyce B. is a nulliparous 41-year-old-female with no known past medical history who had been in her usual state of good health until approximately a year prior to admission when she began seeing her primary care physician for various complaints. Initially she noted urinary urgency and her primary care physician performed a urinalysis, which was negative, and told her to return to the office if symptoms persisted. Over the next few months her symptoms did continue, and Joyce additionally began to feel bloated, developed pain in her lower abdomen, and occasionally had loose bowel movements. On a return visit to her physician she was told it was probably irritable bowel syndrome (IBS) and was encouraged to make dietary changes such as increasing her ingestion of soluble fiber and taking smaller, more frequent meals. A CT scan of the abdomen was ordered at that time, and her physician stated that he would notify her if there were any abnormalities. When Joyce did not hear back from her primary care physician, she assumed that her CT results were normal. She tried the recommended dietary interventions for several months with no improvement in her symptoms, and noted that she was losing the desire to eat. She felt "full" after taking even small meals and believed that her abdominal girth was increasing.

Joyce presented to the Emergency Department at the time of the current admission after noticing new onset of sharp pain and swelling in her right calf upon awakening in the morning. She denied fevers, chills, sweats, chest pain, shortness of breath, or other problems. Joyce used no medications at home and had no known medical allergies. She had no relevant family history, and had never used tobacco, alcohol, or drugs.

Her physical exam was notable for normal vital signs. She was in no apparent distress. Cardiovascular and pulmonary exam were within normal limits. Her abdominal exam was notable for mild distention and generalized lower abdominal tenderness to deep palpation. Her right calf was notably more swollen than her left, and although there was no localized erythema, it was tender to palpation. Laboratory evaluation was unremarkable. Venous Doppler examination of the lower extremities demonstrated bilateral deep venous thrombosis (DVT). CT of the abdomen and pelvis demonstrated a complex, calcified right adnexal mass with adherence to the omentum, regional lymphadenopathy, and a moderate amount of pelvic free fluid.

From: *Evidence-Based Medical Ethics*
By: J.E. Snyder and C.C. Gauthier © Humana Press, Totowa, NJ

The Ethical Dilemma

Joyce was hospitalized by the admitting medical team for treatment of her DVT and further evaluation of her right adnexal mass. The team suspected she had an underlying diagnosis of ovarian cancer and the OB-GYN team was consulted for assistance. The team obtained the outpatient records from Joyce's primary care physician as part of their routine data collection. When the prior records were reviewed by the admitting team, they noted that the report of her abdominal CT scan, performed nearly five months prior to admission, indicated evidence of the adnexal mass. The last line of the CT report stated that "Beth, RN" at the primary care physician's office was notified of the findings at the time of the study. As part of her evaluation in the hospital, Joyce underwent surgical biopsy of her adnexal mass and the team confirmed their greatest concern – a diagnosis of ovarian cancer.

> **Questions for thought and discussion:** What are the responsibilities of Joyce's inpatient medical team with regard to revealing the past CT scan results to her? How should this information be presented to her?

> **Questions for thought and discussion:** How should the team involve Joyce's primary care physician in this process? How about the radiologist that read the CT scan?

The Medicine

According to data from the 2003 report of the United States Cancer Statistics Working Group, ovarian cancer is the seventh most common cancer and the fifth most common cause of cancer death among women in the United States. The U.S. Preventive Services Task Force does not currently recommend any particular routine screening test for ovarian cancer in asymptomatic patients due to the lack of demonstrated benefit in reducing death rates from the disease, when compared to potential harms from the existing screening methods. Despite the past belief that symptom clusters do not give early warning signs of an ovarian cancer diagnosis, recent studies have shown that the vast majority of women with the diagnosis did in fact have preceding symptoms. Most commonly women note abdominal, gastrointestinal, constitutional, pelvic, or urinary symptoms, and have these complaints in both the early and late stages of the disease. Since these symptoms may be common and appear nonspecific, delays in diagnosis often occur. Hence, timely recognition of ovarian cancer symptoms in the primary care setting may lead to earlier diagnostic testing, OB-GYN referral, and treatment – and potentially improved survival rates from the disease.

The Law

Joyce was not informed by her primary care physician that her original CT scan showed an adnexal mass, so she was never referred to a specialist for further evaluation. This suggests that Joyce may have a case of medical malpractice against her physician. Medical malpractice occurs when a medical professional has caused harm or damage to a patient due to negligence in the diagnosis, treatment, or management of a disease or illness. In order for negligence to be "actionable" or viable as a legal case, four conditions must generally be satisfied. First, there must be a legal duty owed to the patient, found in the relevant standard of care; second, there must be a breach of that duty, such that the practitioner failed to conform to the standard of care; third, the patient must have suffered some harm, damage, or loss; and fourth, the breach must have been the "proximate cause" of the harm, damage, or loss.

In this case there was a duty to inform Joyce of the CT results as soon as possible, and to refer her for further evaluation, based on a standard of care for all primary care physicians. Joyce's physician breached that duty by not informing Joyce of the CT results, indicating the presence of an adnexal mass, and by not making the necessary referral. The delay in diagnosis and treatment of Joyce's ovarian cancer may mean that the cancer will be more difficult to treat. If her cancer is not able to be treated, Joyce will suffer the harm of an earlier death and the loss of years of life. Finally, the most difficult aspect of medical malpractice to prove is causation. In this case it would have to be proven that the physician's failure to notify Joyce and make the relevant referral led to the delay in diagnosis and treatment and, ultimately, to her earlier death from ovarian cancer. In other words, if she had been notified five months earlier and referred for treatment, she would have lived longer.

Joyce may not be interested in pursuing a medical malpractice suit while she is undergoing treatment for ovarian cancer and fighting for her life. However, if she dies of the disease, her family may decide to pursue the case. If a lawyer becomes involved, he or she will need to obtain a "certificate of merit," in which another physician certifies that the primary care physician in this case violated acceptable medical practice and this resulted in Joyce's earlier death.

One question that may be important to Joyce and her family, even if a case of medical malpractice is never initiated, is the question of responsibility. Who is responsible for the fact that Joyce was not informed about the adnexal mass and was not referred for further evaluation to ensure an earlier diagnosis and treatment for her ovarian cancer? Beth, the RN at the primary care physician's office, apparently was notified of the results as soon as they were available. Did she inform the physician of the results? If not, why? What was done with Joyce's CT report once it was received in the office? Did the physician ever review the report?

Even if Beth failed in her responsibility to pass on the CT findings at the time of the study, the physician also failed in his responsibility to review the CT report once it was received in his office. Since he ordered the scan and told Joyce he would let her know about any abnormalities, he had a responsibility to follow up by reading the report and meeting with Joyce about the results. If the primary care physician wants to place full responsibility on Beth because she did not inform him of the CT results when she received them, he still will not be able to avoid legal responsibility in a medical malpractice case. According to the legal doctrine of "respondeat superior," the physician may be responsible for his employee's negligence, if that negligence is found to be "actionable".

The Ethics

The medical team should contact Joyce's primary care physician regarding her latest CT and surgical biopsy results. The primary care physician might also be asked why Joyce was not informed of the earlier CT findings. Even if he is not able or willing to discuss this with the medical team, the primary care physician should be encouraged to meet with Joyce to explain why she was not given the results of the earlier scan.

The radiologist would not need to be involved at this point because this specialist appears to have made no mistakes in Joyce's case. The CT results were provided to the primary care physician's office immediately and, since the same results were found five months later, the reading of the original CT scan appears to have been accurate.

Based on the *Principle of Veracity* a member of the medical team needs to inform Joyce about the team's discovery when they reviewed her medical records, including the earlier CT report. They can explain that reviewing her medical records was part of the routine data collection necessary for the further evaluation of her CT results. Joyce needs as much information as the medical team has about what happened in the primary care physician's office to delay her diagnosis so she can decide whether or not to continue as a patient with this doctor, and whether or not to contact a lawyer about pursuing a medical malpractice suit.

The *Principle of Respect for Dignity* also supports providing Joyce with the truth for her emotional well-being. She may feel that her symptoms were ignored or not taken seriously by her primary care physician, and that her trust has been betrayed. The primary care physician can respect Joyce's dignity, and act on the *Principle of Veracity* and the *Principle of Beneficence*, by being honest about the mistakes that were made, offering a sincere apology, and asking for her forgiveness. Joyce needs to have the truth for the reasons noted above. In addition, she may respond better to the cancer treatment if she is not also suffering emotionally trying to understand why her trusted physician treated her this way.

Both Joyce and her primary care physician may benefit from efforts to repair their relationship through a process of reconciliation. This process begins with the practitioner acknowledging the mistake, explaining what happened, and taking personal responsibility for the error. The next step involves apologizing and expressing remorse to the patient, as well as covering the costs of treating injuries resulting from the error, when appropriate. In the final step, the patient may come to forgive the physician and a reconciliation of their relationship may be possible.

The Formulation

Now that the evidence-based medicine, legal precedent, and relevant ethical principles for this case have been reviewed, formulate a strategy to address the ethical conflicts in this case. If necessary, perform additional research into local and state laws and hospital regulations. Consider delving further into the background medical literature to assist with making sound therapeutic decisions. Devise a treatment approach that addresses the needs of the patient and her family, that is both ethically and medically sound, and that is culturally competent. Ensure that the strategy employs fair and appropriate utilization of medical resources, and that the approach is practical and feasible within the limits of the medical system at large. Work out a clear and professional way to communicate the proposal to the patient and her family. Attempt to foresee challenges that may arise in conveying or implementing the plan. Determine what follow-up will be necessary to ensure that the chosen strategy remains successful for the patient in the long-term. Reflect on how the knowledge and skills learned from this case can be used to improve the care of patients that may be encountered in future practice.

Afterthoughts

In this case a patient's diagnosis of ovarian cancer was delayed by several months, and a report from an earlier CT scan suggesting the presence of ovarian cancer was accidentally missed, adding significant time to the delay. The physicians ultimately diagnosing the cancer were obliged to share this information with the patient, creating an uncomfortable position for them and an ethical dilemma as to how best to approach the matter.

The financial burden of liberal – and perhaps mostly unnecessary – testing on our nation's health care system often influences providers to recommend a more conservative approach to patients with mild and nonspecific symptoms. How much delay is acceptable in order to provide cost-effective care without missing the timely diagnosis of a potentially serious disease? Should all patients with nonspecific symptoms immediately undergo diagnostic studies? In the reality of practice environments today, and an increasingly litigious society, how much does fear of missing a diagnosis and getting sued by patients influence a health care practitioner's decision-making process with such patients?

Annotated References/Further Information

Code of Medical Ethics of the American Medical Association: Current Opinions with Annotations, 2006–2007 Edition. Council on Ethical and Judicial Affairs. Annotations prepared by the Southern Illinois University Schools of Medicine and Law.

Snyder L, JD, and Leffler C, JD. Ethics Manual, Fifth Edition. Ethics and Human Rights Committee, American College of Physicians. Annals of Internal Medicine. 142(7):560–582, 5 April 2005.

U.S. Cancer Statistics Working Group. United States cancer statistics: 2003 incidence and mortality. Atlanta (GA): Department of Health and Human Services, Centers for Disease Control and Prevention, and National Cancer Institute; 2007.

U.S. Preventive Services Task Force. Screening for ovarian cancer. Recommendations and rationale. Rockville, MD: Agency for Health Care Research and Quality 2004.

Goff BA et al. Frequency of symptoms of ovarian cancer in women presenting to primary care clinics. JAMA. 291(22), 9 June 2004, pp. 2705–2712.

Goff BA et al. Ovarian cancer diagnosis: results of a national ovarian cancer survey. Cancer. 2000;89:2068–2075.

Berlinger N. " 'Missing the mark': medical error, forgiveness, and justice," pp. 119–134, in Accountability: Patient Safety and Policy Reform, Virginia A. Sharpe, editor, Washington, DC: Georgetown University Press, 2004.

Kohn LT, Corrigan JM and Donaldson MS, editors. To err is human: building a safer health system, Institute of Medicine, Washington, DC: National Academy Press, 2000.

The following website provides a useful introduction to medical malpractice:
http://injury.findlaw.com/medical-malpractice/. Accessed October 15, 2007.

Case 18
When a Patient Codes

The Patient

Debra W. is a 45-year-old female with no known past medical history who initially noted left leg pain on the evening prior to admission. She took some ibuprofen and went to bed, sleeping through the night. Upon awakening and rising, she developed sharp, constant chest pain, worse with inspiration, and associated with significant shortness of breath. She alerted her husband, Carl, and 23-year-old son, Gregory, to call for paramedics, and was brought to the Emergency Department. Upon presentation, she was notably uncomfortable and tachypneic, and too short of breath to speak in full sentences. Her physical exam was notable for a heart rate of 120 beats per minute, blood pressure of 90/49, and an oxygen saturation of 82 percent on room air. Additionally she had a swollen and exquisitely tender left calf. Arterial blood gas testing was consistent with respiratory alkalosis and hypoxemia. An electrocardiogram showed a right bundle branch block. Laboratory evaluation was notable for mildly elevated troponin I and brain natriuretic peptide (BNP) levels. CT angiogram of her chest demonstrated large bilateral pulmonary emboli. Stat echocardiography was performed and showed evidence of right ventricular dysfunction.

The Ethical Dilemma

Debra was admitted to the intensive care unit, and she and her family were informed of the serious nature of her diagnosis. Based on the severity of her embolic burden, thrombolytic therapy was ordered. However, before this could be started, she had an asystolic arrest and resuscitation was begun. The resident leading the code asked Debra's family to step out of the room, but the ICU nurse intervened and asked "shouldn't they be allowed to stay?"

Questions for thought and discussion: Should families be present during resuscitation attempts on their loved one? What are the advantages and disadvantages of having them present?

From: *Evidence-Based Medical Ethics*
By: J.E. Snyder and C.C. Gauthier © Humana Press, Totowa, NJ

The Medicine

Venous thromboembolism is a common and deadly disease. Almost 25 percent of patients with pulmonary embolism present with sudden death. In the remaining patients, factors such as right ventricular dysfunction and hypotension predict poor long-term survival and acute mortality remains around 65 percent. Thrombolytic therapy should be considered in patients with these ominous signs.

The Law

Suggesting that a patient's family members be present during a resuscitation attempt is a fairly recent development. Currently, there are no laws or court cases that would either require hospitals to permit family members to be present during a code or prohibit hospitals from allowing this. The American College of Physicians includes the following statement in their 2005 Ethics Manual: "When clinicians perform cardiopulmonary resuscitation, family members should usually have the choice to be present." There is mounting support for this concept, particularly among nursing organizations. The Emergency Nurses Association and American Association of Critical Care Nurses both have crafted position statements in this regard, and the American Heart Association endorses this position as well. A number of surveys indicate that most families want to be present during codes, and that health care providers who have had experiences with this concept support the idea as well. Although this remains a controversial subject, the concept of allowing loved ones to be together at the end of life is a valid one to be further studied.

The Ethics

A patient's bedside, when resuscitation is being initiated, is not the place and time to introduce the issue of family presence during a code. The ICU nurse should not have intervened at this point, since there was no time for a reasoned discussion of this issue among the care providers or with the family. It would have been better for the nurse to broach the topic with the director of nursing who could discuss it with the appropriate hospital administrators. The hospital may want to develop a policy about the presence of family members at resuscitation attempts.

If the hospital chooses to allow this, the policy should include asking patients, if possible, whether or not they want members of their family to attend a code and asking family members whether or not they want to be present. This could be included in the discussion of code status for each patient. If the patient decides to be "full code," then questions about family presence would be appropriate. There should probably be an age requirement for family members allowed to witness a resuscitation attempt, such as 18- or 21- years-of-age. In addition, before family members are asked to make a decision about attending a code, they should be educated about the interventions that will be employed and the possible outcomes.

The policy should also include procedures to protect the patient from any disturbances caused by family presence during resuscitation. This might include guidelines for assigning a nurse or a member of the support staff to be with the family and for the gentle removal of family members who become overly distraught or disruptive.

When deciding whether or not to have a policy allowing family members the choice of being present at resuscitation, hospital administrators and medical staff could consider a variety of ethical principles and arguments. Applying the *Principles of Beneficence* and *Non-Maleficence*, the good promoted for the patient should be compared with the possibility of harm. Many patients will be unconscious and unaware of who is there during the resuscitation attempt. However, some patients may be aware of their surroundings and may be comforted by the presence of family members. Notice, too, that as long as the relevant policy is formulated to protect the smooth functioning of the resuscitation process, there is little harm that could come to the patient from having family members present during a code.

Both the *Principle of Respect for Autonomy* and the *Principle of Respect for Dignity* would support a policy that allows patients, if possible, to decide whether or not their family members could be present during a code. With this policy the patient would be permitted to accept or reject resuscitation with family members present, a treatment option that is being offered by the hospital. This policy would also demonstrate respect for the patient's emotions and important relationships. The *Principle of Respect for Dignity* also supports a policy that allows family members the choice of witnessing the resuscitation of their loved one. In this way the emotions of family members and their relationships with the patient are recognized and respected.

There are a number of additional arguments that could be used to support family members' attendance at the patient's code. This would allow family members who have been demanding resuscitation against the recommendation of the medical team to experience the reality of a code. When resuscitation is considered to be futile or inappropriate, actually seeing what happens may persuade family members that it is not in the patient's best interests to be resuscitated in the future. Family members who are permitted to witness a code may feel that "everything possible" was done and may be less likely to blame care providers if the code is not successful. This policy also allows family members to be with and comfort the patient at what may be the end of the patient's life. Finally, attending the resuscitation may also help family members with the grieving process if the patient does not survive.

There are also arguments against this kind of policy. Seeing what is done to the patient during a resuscitation attempt may be traumatic for family members. They may become so distraught that they disrupt the resuscitation efforts. If the code is unsuccessful, this may leave family members with a painful and unpleasant last memory of their loved one.

These arguments will not be very persuasive against a policy that permits adult family members to make this decision, based on truthful information about what will happen during the code and a realistic understanding of the possible outcomes, and also includes safeguards to protect the integrity of the resuscitation process. Capable adults should be allowed to make their own decisions about their experience of a family member's death, just as capable adult patients should be permitted to make their own decisions about recommended medical treatment at the end of their lives.

The Formulation

Now that the evidence-based medicine, legal precedent, and relevant ethical principles for this case have been reviewed, formulate a strategy to address the ethical conflicts in this case. If necessary, perform additional research into local and state laws and hospital regulations. Consider delving further into the background medical literature to assist with making sound therapeutic decisions. Devise a treatment approach that addresses the needs of the patient and her family, that is both ethically and medically sound, and that is culturally competent. Ensure that the strategy employs fair and appropriate utilization of medical resources, and that the approach is practical and feasible within the limits of the medical system at large. Work out a clear and professional way to communicate the proposal to the patient and her family. Attempt to foresee challenges that may arise in conveying or implementing the plan. Determine what follow-up will be necessary to ensure that the chosen strategy remains successful for the patient in the long-term. Reflect on how the knowledge and skills learned from this case can be used to improve the care of patients that may be encountered in future practice.

Afterthoughts

In this case an ethical dilemma occurred when a health care team was determining if a family should remain present during the resuscitation of a patient. The patient in question was young and had a 23–year-old son who witnessed her arrest. How would this case have been different if her son had been younger? At what age does someone have the emotional strength to handle the stress of witnessing the resuscitation of a loved one? Is 18-years-of-age old enough? If family should be present, should other close people in the patient's life? How many people present is too many? Does family presence at a resuscitation make it more difficult for health care providers to perform their job well? Should providers be concerned that failed attempts at life-saving procedures (e.g., intubation) that are witnessed by family members could lead to blame or even litigation?

Annotated References/Further Information

Code of Medical Ethics of the American Medical Association: Current Opinions with Annotations, 2006–2007 Edition. Council on Ethical and Judicial Affairs. Annotations prepared by the Southern Illinois University Schools of Medicine and Law.

Snyder L, JD, and Leffler C, JD. Ethics Manual, Fifth Edition. Ethics and Human Rights Committee, American College of Physicians. Annals of Internal Medicine. 142(7):560–582, 5 April 2005. The American College of Physicians recommends that family members be given the choice of being present at resuscitation in the subsection, Patients Near the End of Life.

Heit JA. The epidemiology of venous thromboembolism in the community: implications for prevention and management. Journal of Thrombosis and Thrombolysis. 21(1):23, 2006.

Konstantinides S. Pulmonary embolism: impact of right ventricular dysfunction. Current Opinion in Cardiology. 20:496, 2005.

Davidson JE. Family presence at resuscitation: What if? Critical Care Medicine. 34(12):3041, 2006.

Mazer MA et al. The public's attitude and perception concerning witnessed cardiopulmonary resuscitation. Critical Care Medicine. 34(12):2925, 2006.

Case 19
When a Patient Places a Practitioner at Risk

The Patient

Kevin B. is a 29-year-old male with a past history of cocaine and alcohol abuse who presented to the Emergency Department after having an altercation in a bar and sustaining a laceration to his left cheekbone. He had no other known medical history and took no home medications. Although he reported active use of crack cocaine, Kevin had also used intravenous heroin in the past. He generally consumed four to six alcoholic beverages on a daily basis. He reported no past testing for HIV or viral hepatitis, and refused testing at this time.

On physical exam he was a well developed, well nourished male in no apparent distress. There was an open 4 cm laceration on the left face, extending to about 2 cm below the eye. Other than the presence of several tattoos, the remainder of his physical exam was within normal limits.

The Ethical Dilemma

Due to the size and location of Kevin's laceration, the emergency physician called the on-call surgical resident to suture and dress the wound. Informed consent was obtained, and the resident proceeded to clean the wound and suture it together. In the process, the resident sustained a needlestick injury that penetrated through his surgical gloves and into his skin, drawing a scant amount of blood. He immediately stopped the procedure, removed his gloves and thoroughly rinsed the puncture wound under the sink with soap and water. He then asked Kevin to give consent for HIV and viral hepatitis testing, but Kevin again refused, stating "I don't have those." When the surgical resident pleaded with him to give consent for the sake of his own possible exposure, Kevin replied, "that's not my problem."

Question for thought and discussion: Can Kevin be tested for HIV without consent in this case? How about for viral hepatitis?

From: *Evidence-Based Medical Ethics*
By: J.E. Snyder and C.C. Gauthier © Humana Press, Totowa, NJ

Question for thought and discussion: If Kevin refuses HIV testing, can a surrogate test, such as a CD4+-lymphocyte count, be ordered without his consent?

The Medicine

Needlestick injuries in health care personnel are common, with a frequency of greater than 600,000 incidents per year in the United States. It is also estimated that only half of the true total number of events are reported. Surgical residents are at particular risk for this injury for many reasons. They are continuously exposed to sharp instruments and blood while trying to develop their surgical skills. Data also suggests that their patient population has high prevalence rates of bloodborne viruses such as hepatitis B (HBV), hepatitis C (HCV), and human immunodeficiency virus (HIV).

If a health care provider is exposed to a patient's blood through a needlestick, they should immediately and thoroughly cleanse the area with soap and water and then notify their facility's infection control department for advice regarding postexposure prophylaxis (PEP) to reduce HBV and HIV transmission. Currently, no PEP exists to prevent hepatitis C transmission.

After an occupational exposure with a single needlestick, the risk of hepatitis B transmission is extremely low for those health care providers who have been vaccinated for the virus. Without vaccination the risk of infection ranges from 6 to 30 percent. Risks for hepatitis C and HIV infection are 1.8 percent and 0.3 percent, respectively. Between 1985 and 2001 the CDC reported a total of 57 documented and 138 possible cases of HIV transmission to health care personnel via occupational exposure.

CD4+-lymphocyte counts, from purely a medical standpoint, are not necessarily an accurate surrogate marker for diagnosing HIV infection. Although the CD4+-lymphocyte count must be monitored in patients with known HIV disease to assess disease progression, there are persons living with HIV that do not have significantly suppressed CD4+-lymphocyte counts. These so-called long-term non-progressors (LTNP), accounting for approximately 1 to 5 percent of those infected with HIV-1, remain asymptomatic, have low or undetectable viral loads and maintain CD4+-lymphocyte counts $>500/\mu L$ for several years without the use of antiretroviral medications. These individuals are of great research interest for obvious reasons.

Additionally, there are persons with low CD4+-lymphocyte counts who do not have HIV infection. One study of sexually active, heterosexual women in New York, who did not have a history of intravenous drug use and were HIV-1 negative, estimated that, at any given time, up to 4.1 percent may have a transient absolute CD4+-lymphocyte count less than $300/\mu L$. Aside from HIV infection, lymphocytopenia may be due to other autoimmune diseases, infections, medications, and lymphoma. Idiopathic CD4+ lymphocytopenia (ICL) has also been described.

The Law

Informed consent is not required to test a patient for viral hepatitis in any state. With the exception of HIV tests and certain genetic assays, most patient testing falls into the blanket category of general consent for medical care, and this is usually included in a document that a patient signs when entering a health care facility for evaluation and treatment. Thirty-two states legally require verbal or written informed consent from a patient prior to HIV testing (see Table 19.1), and almost all states have some form of mandatory pre-test counseling required. However, there are currently no laws

Table 19.1 States with laws requiring written or verbal informed consent prior to testing a patient for HIV infection

Alabama
Arizona
California
Colorado
Connecticut
Delaware
District of Columbia
Florida
Hawaii
Illinois
Kentucky
Louisiana
Maine
Maryland
Massachusetts
Michigan
Montana
New Hampshire
New Jersey
New Mexico
New York
North Dakota
Ohio
Oregon
Pennsylvania
Rhode Island
Tennessee
Texas
Vermont
Washington
West Virginia
Wisconsin

Source adapted from: Hodge JG. Advancing HIV prevention initiative - a limited legal analysis of state HIV statutes. Initial Assessment for the Centers for Disease Control and Prevention. As of September 30, 2004. Acquired from: *http://www.publichealthlaw.net/Research/PDF/ AHP%20Report%20-%20Hodge.pdf.* Accessed October 15, 2007.

that specifically call for informed consent before performing surrogate tests on a patient, such as CD4+-lymphocyte counts or HIV viral loads. Circumventing consent laws by using such assays may be considered by many to be deceitful, but legal precedent against this practice has not yet been established. Many states have regulations allowing caregivers a waiver to test individuals for HIV without obtaining their consent, but generally this is reserved for circumstances when the patient does not have the capacity to provide consent (see Table 19.2).

Based on CDC recommendations for national surveillance of HIV infection, 47 states and the District of Columbia report new diagnoses to the CDC using a confidential, name-based system. The three other states (Hawaii, Maryland, and Vermont) use a confidential code-based system. It is important to note that, in 42 states, there are also regulations mandating the reporting of CD4+-lymphocyte count results, and 43 states require the reporting of HIV viral load test results (see Table 19.3). This

Table 19.2 States with laws allowing non-consented testing of HIV in patients without the capacity to provide consent

States permitting non-consented HIV testing if the treating physician believes that test results directly and immediately will improve the patient's medical care:
Alabama
Arizona
Arkansas
Colorado
Connecticut
Delaware
Hawaii
Illinois
Indiana
Kentucky
Mississippi
New Hampshire
Utah

States permitting non-consented HIV testing in emergency situations:
Georgia
Pennsylvania
New Mexico
Ohio
West Virginia
Wisconsin

States permitting non-consented HIV testing if a surrogate individual provides consent:
California
Florida
Maryland
Montana
New York
North Carolina
North Dakota
Rhode Island

Source adapted from: Halpern SD. HIV testing without consent in critically ill patients. JAMA. 2005;294:734–737.

Table 19.3 States requiring reporting of CD4+-lymphocyte count and HIV viral load results by the testing laboratory

States requiring CD4+-lymphocyte count results to be reported:
Alabama
Alaska
Arizona
Arkansas
Colorado
Connecticut
Delaware
District of Columbia
Florida
Hawaii
Idaho
Illinois
Indiana
Iowa
Kansas
Kentucky
Louisiana
Maine
Maryland
Massachusetts
Minnesota
Mississippi
Missouri
Nebraska
Nevada
New Hampshire
New Jersey
New Mexico
New York
North Dakota
Ohio
Oklahoma
Oregon
Pennsylvania
Puerto Rico
Rhode Island
South Carolina
Tennessee
Texas
Utah
Washington
West Virginia
Wisconsin
Wyoming

States requiring HIV viral load results to be reported:
Alaska
Arizona
Arkansas
California

(continued)

Table 19.3 (continued)

Colorado
Delaware
District of Columbia
Hawaii
Idaho
Illinois
Indiana
Iowa
Kansas
Kentucky
Louisiana
Maine
Maryland
Minnesota
Mississippi
Missouri
Montana
Nebraska
Nevada
New Hampshire
New Jersey
New Mexico
New York
North Carolina
North Dakota
Ohio
Oklahoma
Oregon
Pennsylvania
Puerto Rico
Rhode Island
South Carolina
Tennessee
Texas
Utah
Vermont
Virginia
Washington
West Virginia
Wisconsin
Wyoming

Source adapted from: CDC (*www.cdc.org*). State listings as of
January 2005. Threshold laboratory values for which reporting is
mandated varies with individual states; refer to the CDC website
for further details.

reporting is done regardless of whether the patient's HIV status is known, with the
purpose of improving surveillance of HIV-related illness. Therefore, someone who
has not provided informed consent for HIV testing, but is screened by a caregiver
with a test such as a CD4+-lymphocyte count without their knowledge or consent,

may still have this laboratory information reported to a national, government-associated database.

The Ethics

The surgical team should use the *Principle of Beneficence* and the *Principle of Non-Maleficence* to determine if HIV or other surrogate testing would be in Kevin's best interests. It could be argued that, if Kevin does have HIV, it is better for him to know this as soon as possible so that he can begin treatment. If the test is positive, the practitioners could prevent the harm to Kevin that would come from being infected with HIV and going without treatment. This is particularly important now that antiviral drugs have proven to be successful in prolonging the lives of those with HIV. However, the surgical team must also consider the possibility that testing Kevin could cause him harm. Because the confidentiality of positive test results is protected under state laws, it is unlikely that any harm would come to Kevin, even if he tests positive.

Considering the harm that could be prevented, and the fact that no harm would be done by the tests, it appears that ordering an HIV or surrogate test for Kevin would be in his best interests. This would be true even if Kevin's health is not the primary motivation for ordering the HIV or surrogate test. In fact, because this is not the primary motivation, it is especially important that these principles are used to determine the impact on Kevin's interests if any of these tests are done.

According to the *Principle of Respect for Autonomy*, Kevin would normally be permitted to refuse the HIV test as well as the surrogate tests. However, like all of the principles of medical ethics, this is not an absolute principle. It may be out-weighed by more important considerations in a moral dilemma. In this case, the requirement to allow patients to accept or refuse medical interventions is out-weighed by the importance of preventing harm to the surgical resident from having been infected with the AIDS virus and not getting treatment.

We can justify overriding the *Principle of Respect for Autonomy* by comparing what is lost when a patient's autonomy is violated and what is lost when it is honored. If Kevin is tested for HIV, against his wishes, he will lose some of his privacy in the sense of control over information about his medical status. If he tests positive he will be informed of this and several other people will have this information, including the surgical resident. On the other hand, if Kevin's autonomy is honored and he is not tested, the surgical resident will not know his own risk for HIV infection and Kevin, if he is infected with the virus, will lose the opportunity to begin antiviral drugs for HIV.

Based on these ethical principles it would be morally justified to order the HIV test without Kevin's consent, even though it is illegal. It is somewhat surprising that state laws do not include testing after accidental exposure to a patient's blood as an exception to the informed consent requirement for HIV testing. This seems obvious, given the success of antiviral drugs in combating HIV. In the case of accidental exposure to HIV, as in this case, practitioners should check with hospital legal counsel to determine the current law regarding HIV testing in their individual states, as these laws may change over time.

Because it is illegal to perform the HIV test against Kevin's wishes, it would be better to order one of the surrogate tests that can be done without consent, if the surgical residents and faculty believe it would be helpful to the resident involved in this case. It will be important, however, to inform Kevin that the test will be done and the reasons for it and to provide him with the test results when they are available. If the results indicate that Kevin may be infected with HIV, he should also receive counseling and a referral for follow-up medical care. It would only be deceptive if the HIV test or surrogate tests were done without Kevin's knowledge.

The *Principle of Respect for Dignity* includes the requirement to respect the confidentiality of a patient's medical condition, including the results of medical tests. However, confidentiality may be overridden by the need to prevent harm to others. This is a morally justified exception to medical confidentiality and is recognized by law as well. This explains why positive HIV tests, as well as positive tests for other communicable diseases, are reportable to the Public Health Department.

The Formulation

Now that the evidence-based medicine, legal precedent, and relevant ethical principles for this case have been reviewed, formulate a strategy to address the ethical conflicts in this case. If necessary, perform additional research into local and state laws and hospital regulations. Consider delving further into the background medical literature to assist with making sound therapeutic decisions. Devise a treatment approach that addresses the needs of the patient and his family, that is both ethically and medically sound, and that is culturally competent. Ensure that the strategy employs fair and appropriate utilization of medical resources, and that the approach is practical and feasible within the limits of the medical system at large. Work out a clear and professional way to communicate the proposal to the patient and his family. Attempt to foresee challenges that may arise in conveying or implementing the plan. Determine what follow-up will be necessary to ensure that the chosen strategy remains successful for the patient in the long-term. Reflect on how the knowledge and skills learned from this case can be used to improve the care of patients that may be encountered in future practice.

Afterthoughts

In this case a patient declined to consent for HIV testing, even after their physician was injured by a needlestick during their care. Why is general consent for medical evaluation and treatment inadequate to cover consent for HIV testing and certain genetic assays? Is the stigma of HIV infection as relevant an argument now for special testing status as it was when the virus was newly discovered? Should testing for all reportable conditions require informed consent?

Annotated References/Further Information

Code of Medical Ethics of the American Medical Association: Current Opinions with Annotations, 2006–2007 Edition. Council on Ethical and Judicial Affairs. Annotations prepared by the Southern Illinois University Schools of Medicine and Law.

Snyder L, JD, and Leffler C, JD. Ethics Manual, Fifth Edition. Ethics and Human Rights Committee, American College of Physicians. Annals of Internal Medicine. 142(7):560–582, 5 April 2005.

Exposure to blood:what health care personnel need to know. Information from the Centers for Disease Control and Prevention, National Center for Infectious Diseases, Division of Health Care Quality Promotion and Division of Viral Hepatitis. July 2003. Obtained from *http://www. cdc.gov/ncidod/dhqp/pdf/bbp/Exp_to_Blood.pdf*. Accessed October 15, 2007.

NIOSH alert: preventing needlestick injuries in health care settings. Washington, DC: National Institute for Occupational Safety and Health, 1999 (Publication no. 2000–108). Obtained from: http://www.cdc.gov/niosh/2000–108.html. Accessed October 15, 2007.

Makary MA et al. Needlestick injuries among surgeons in training. New England Journal of Medicine. 356(26):2693–2699.

Rodés B et al. Differences in disease progression in a cohort of long-term non-progressors after more than 16 years of HIV-1 infection. AIDS. 18(8):1109–16, 2004 May 21.

DeHovitz JA et al. Idiopathic CD4+ T-Lymphocytopenia. New England Journal of Medicine. 329(14):1045–1046.

Walker UA and Warnatz K. Idiopathic CD4 lymphocytopenia. Current Opinion in Rheumatology. 18:389–395.

Klevens R et al. Impact of laboratory-initiated reporting of CD4+ T Lymphocytes on U.S. AIDS surveillance. Journal of Acquired Immune Deficiency Syndromes & Human Retrovirology. Volume 14(1), 1 January 1997, pp. 56–60.

http://www.cdc.gov/hiv/topics/surveillance/reporting.htm. Accessed October 15, 2007.

http://www.cdc.gov/hiv/topics/surveillance/resources/reports/2005supp_vol11no2/pdf/table14. pdf. Accessed October 15, 2007.

Halpern SD. HIV testing without consent in critically ill patients. JAMA. 2005;294:734–737.

Hodge JG. Advancing HIV prevention initiative - a limited legal analysis of state HIV statutes. Initial Assessment for the Centers for Disease Control and Prevention. As of September 30, 2004. Acquired from:

http://www.publichealthlaw.net/Research/PDF/AHP%20Report%20-%20Hodge.pdf. Accessed October 15, 2007.

http://www.edhivtestguide.org/EDHILega-4071.html. Accessed October 15, 2007.

Case 20
When Patient Non-Adherence Dictates Therapeutic Options

The Patient

Samantha R. is a 32-year-old female with a past medical history of end-stage renal disease from post-streptococcal rapidly progressive glomerulonephritis (RPGN) at age 19, who received a cadaveric renal transplant at age 24. At 28-years-old, her fiancé was killed in a motorcycle accident. After this event, she became depressed and stopped taking her immunosuppressant medications. Subsequently, her transplant failed and she has since been on hemodialysis. Her depression was well-controlled after these events on one medication, and this treatment was eventually discontinued altogether with no further episodes of depression. She has been adherent with all medical recommendations and treatment since the transplant failure.

At her most recent dialysis treatment she complained of fevers and rigors at home. Blood cultures were drawn and it was found that Samantha had gram-negative rod bacteremia. She was directly admitted to the hospital for intravenous antibiotic therapy. Physical exam was unremarkable on admission. Laboratory work-up was notable only for her renal disease and bacteremia. Chest radiography and electrocardiogram were within normal limits.

The Ethical Dilemma

Samantha was initially started on broad spectrum antibiotics, which were then changed to a narrower spectrum choice based on further blood culture identification and sensitivity data. She had excellent response to treatment and her hospital course was otherwise uncomplicated. Prior to discharge, she told her primary physician that she was tired of the limitations of life on hemodialysis, such as the treatment schedule, the symptoms of fatigue, and the complications such as bacteremia. She stated that her sister was willing to be a living organ donor for her so that she could have a second chance at transplantation.

Questions for thought and discussion: Can Samantha be a transplant recipient a second time, after non-adherence to therapy caused failure of her first

From: *Evidence-Based Medical Ethics*
By: J.E. Snyder and C.C. Gauthier © Humana Press, Totowa, NJ

transplant? What if the episode of non-adherence was due to a reversible and theoretically non-recurring cause, such as grief reaction?

The Medicine

End-stage renal disease (ESRD) is a common problem in the United States, with a point prevalence of 472,099 patients under treatment for this condition as of December 31, 2004. The most common causes are diabetes and hypertension, but glomerulonephritis and cystic kidney disease contribute the majority of the remaining cases. In 2004, a total of 335,963 patients in the United States received dialysis and 16,905 kidney transplants were performed. Of the transplants, approximately 60.5 percent were cadaveric, 25.6 percent were from a living related donor and 13.7 percent were from a living unrelated donor. Based on Organ Procurement and Transplantation Network (OPTN) data, as of August 20, 2007, there are 72,820 active waiting list candidates for a renal transplant. The median wait time for patients to receive a donor kidney is 38.8 months, according to five years of data from the Scientific Registry of Transplant Recipients. Cadaveric renal transplants generally have an average lifespan of eight years, whereas living donor kidneys last about 11 years. The 10-year survival rates for patients on dialysis, following cadaveric renal transplantation, and following living donor renal transplantation are 10 percent, 59.4 percent and 75.6 percent, respectively. According to 2004 data the total Medicare cost for patients undergoing dialysis therapy that year was $16.3 billion, and just over $1 billion was spent on costs related to renal transplantation. The costs per-patient-per-year, however, are greater for transplant patients than dialysis patients – $99,000 versus $66,650.

Although receiving a renal transplant mandates that a patient take immunosuppressant medications daily for the life of the transplanted organ, most patients prefer the freedom of having a transplant to the requirements of ongoing dialysis therapy and the symptoms and dietary restrictions of living with ESRD. Due to the large number of patients awaiting donor kidneys and the long wait times associated with receiving one, transplant centers do rigorous evaluations before accepting a candidate onto their recipient waiting list. In addition to thorough medical evaluations to ensure that the transplant surgery will be tolerated, patients must meet certain other criteria, established to help ensure the highest likelihood of graft success. Although standards may vary at different centers, a general consensus is that the patient may not have advanced cardiac or pulmonary disease, severe neurologic or psychiatric illness, metastatic cancer, active infection, ongoing substance abuse, or a history of non-adherence to medical therapy. Historically, centers have varied on their willingness to transplant solid organs to patients that are HIV-positive, although recent data from Qiu, et al. suggests that five-year patient and transplant survival rates may be similar for patients with and without HIV infection who receive organs from the same donor. HIV seropositivity is generally not considered any longer to be an absolute contraindication to receiving an organ transplant.

Patients with a history of treatable, mild depression are generally not excluded from organ donation waiting lists. However, it is less likely for a patient who rejected a transplanted kidney due to non-adherence with immunosuppressant therapy to be approved for receiving a second transplant. Patients generally undergo the same screening process for transplantation, regardless of the source of the donor organ (i.e., cadaveric or living related donor). Simply having a relative willing to donate a kidney should not make the transplant team more likely to proceed with transplantation as much as the encouraging results of a thorough psychiatric evaluation of this patient would.

The Law

In 1968, the Uniform Anatomical Gift Act was passed by the U.S. Congress allowing individual 18 years or older to donate organs for medical purposes with a simple witnessed document, such as a donor card. The act was adopted by all 50 states and the District of Columbia. The National Organ Transplant Act of 1984 established a grant and contract for an organ procurement and transplantation network, which also expressly prohibited the buying and selling of organs. In 1986 the United Network for Organ Sharing (UNOS) was awarded the federal contract to operate the Organ Procurement and Transplantation Network (OPTN). UNOS registers patients for transplant, matches donors to recipients, and coordinates the organ sharing process.

According to the UNOS website, living related kidney transplants have been done since 1954, with living donors usually related to recipients by blood. Potential living donors must be in good health and between 18 and 60 years of age. They must provide a medical history, have a complete physical exam, and go through a series of screening tests, including tissue typing and blood type compatibility tests. With any living donation, the donor's expenses for "travel, housing, and lost wages" may be covered without violating the prohibition against buying or selling organs (National Organ Transplant Act, section 301).

There is presently no law concerning second transplants. Samantha's primary care physician may send her name and medical record to the transplant center in her area. Each individual transplant center determines who is accepted into the transplant program and registers accepted patients with UNOS, which places accepted patients on the national waiting list. Upon receiving Samantha's sister's medical information, the transplant center will determine if she is a tissue match for Samantha.

The Ethics

Samantha's primary physician should use the *Principle of Beneficence* and the *Principle of Non-Maleficence* to determine whether or not to recommend her for a second kidney transplant. Notice, first, that these two principles are focused exclusively on the patient, so that the sister's best interests are not considered, nor are the interests of the society, when these principles are applied. A second transplant would promote good for Samantha by improving her quality of life in the short-term and, perhaps, by saving her life in the long-term. It would help prevent the harmful consequences of ESRD that Samantha is experiencing now (e.g., infections, fatigue, and limitations of dialysis). The harm that may come to Samantha includes the risks of the transplant surgery and the side effects of the immunosuppressant medications she must take for the rest of her life. Considering these risks and benefits, it could be argued that a second transplant would be in Samantha's best interests.

It might appear that, based on the *Principle of Respect for Autonomy*, Samantha's physician should submit her name to the transplant program for a second transplant simply because she has requested this. However, this principle requires only that capable patients be allowed to accept or refuse recommended medical interventions. The principle will only apply, in this case, if Samantha's physician believes a second transplant would be in her best interests and recommends this to her.

Based on the *Principle of Respect for Dignity*, Samantha's physician may also consider the importance of the family relationships involved, since Samantha's sister seems to want to donate her kidney out of love and concern for Samantha.

According to the *Principle of Distributive Justice*, health care resources should be distributed fairly among the members of society. Organs for transplantation are among the most valuable and scarce of these resources. This raises the question of whether a second kidney transplant for Samantha, after her noncompliance caused the failure of her first transplant, would be a fair distribution of transplantable organs. However, Samantha's physician should not make this determination. This is up to the transplant center and the United Network for Organ Sharing, whose function is to insure the fair and equitable distribution of transplantable organs. Samantha's primary physician should decide whether to recommend her for a second transplant based exclusively on her best interests.

If Samantha is recommended for another transplant, the transplant program will consider her noncompliance, her depression and its cause, as well as her renewed compliance following the successful treatment of her depression. The transplant physicians should place a great deal of weight on the reasons for Samantha's noncompliance with therapy following her first transplant. Her grief after losing a cherished loved one understandably caused a depression that made it difficult for her to comply with post-transplant therapy. She may have felt that she had little reason to go on living. The fact that her depression was successfully treated and that she has been compliant with all medical recommendations since then should indicate a willingness to be compliant in the future. Because her noncompliance can be traced to a treatable depression, with an understandable cause, she may still be a viable candidate for a second transplant.

The Formulation

Now that the evidence-based medicine, legal precedent, and relevant ethical principles for this case have been reviewed, formulate a strategy to address the ethical conflicts in this case. If necessary, perform additional research into local and state laws and hospital regulations. Consider delving further into the background medical literature to assist with making sound therapeutic decisions. Devise a treatment approach that addresses the needs of the patient and her family, that is both ethically and medically sound, and that is culturally competent. Ensure that the strategy employs fair and appropriate utilization of medical resources, and that the approach is practical and feasible within the limits of the medical system at large. Work out a clear and professional way to communicate the proposal to the patient and her family. Attempt to foresee challenges that may arise in conveying or implementing the plan. Determine what follow-up will be necessary to ensure that the chosen strategy remains successful for the patient in the long-term. Reflect on how the knowledge and skills learned from this case can be used to improve the care of patients that may be encountered in future practice.

Afterthoughts

In this case, a patient desired a second chance at receiving a renal transplant after non-adherence with therapy caused the first transplanted organ to fail. Why is it that certain medical interventions, such as organ transplantation and the medical treatment of hepatitis C and HIV, require that a patient prove themselves "reliable" before treatment is given? What is it about these specific interventions that set them apart from other ones that have little or no eligibility criteria to be met? How does one objectively determine that a patient is deserving of the treatment opportunity? Is the placing of limitations on treatment options overly paternalistic?

Annotated References/Further Information

Code of Medical Ethics of the American Medical Association: Current Opinions with Annotations, 2006-2007 Edition. Council on Ethical and Judicial Affairs. Annotations prepared by the Southern Illinois University Schools of Medicine and Law.

Snyder L, JD, and Leffler C, JD. Ethics Manual, Fifth Edition. Ethics and Human Rights Committee, American College of Physicians. Annals of Internal Medicine. 142(7):560–582, 5 April 2005.

http://kidney.niddk.nih.gov/kudiseases/pubs/kustats/index.htm. Accessed October 15, 2007.

The Scientific Registry of Transplant Recipients: http://www.ustransplant.org/csr/current/nats. aspx. Accessed October 15, 2007.

United States Renal Data System. USRDS 2006 Annual Data Report. Bethesda, MD: National Institute of Diabetes and Digestive and Kidney Diseases (NIDDK), National Institutes of Health (NIH), U.S. Department of Health and Human Services (DHHS); 2006. Available at www.usrds.org. Accessed October 15, 2007.

United Network for Organ Sharing. Available at www.unos.org. Accessed August 20, 2007. For updates, call 804–330–8576 or fax 804–323–3794.

http://viper.med.unc.edu/surgery/AbdominalTransplant/kidney-pancreas_surgical_discussion. html#candidates. Accessed October 15, 2007.

Qiu J et al. HIV-positive renal recipients can achieve survival rates similar to those of HIV-negative patients. Transplantation. 81(12):1658–61, 2006 Jun 27.

Gow PJ at al. Solid organ transplantation in patients with HIV infection. Transplantation 2001; 72:177.

http://www.kidney.org.uk/Medical-Info/transplant/txsurvival.html. Accessed October 15, 2007.

The best websites for information on organ donation are: *www.unos.org* and *www.optn.org*. These websites include links on the history of organ donation and, in particular, living donation.

Case 21
When Family Members Limit Caregiver–Patient Communication

The Patient

Andrew N. is a 47-year-old male with a past medical history of alcohol and intravenous drug use, hepatitis C, and cirrhosis. He was brought to the Emergency Department by his parents who had come to his apartment to check on him after not hearing from him for several days. He was in an altered mental state on presentation; he was lethargic, confused, and disoriented. Minimal verbal history could be obtained from Andrew himself. Per his parents, Andrew had an extensive history of heroin addiction and had failed multiple past attempts at rehabilitation. He had not sought medical care otherwise in several years. He took no medications and had no known drug allergies. In addition to his alcohol and drug abuse, Andrew had a 50 pack-year history of tobacco use. He had been previously married, but was divorced and long estranged from his ex-wife. He had no known children.

On physical examination, he had low-grade fever and borderline hypotension. He was cachectic, visibly jaundiced, and had marked icterus. Mucous membranes were very dry. His neck was supple, and there was no lymphadenopathy. Heart and lung exams were unremarkable. His abdomen was mildly distended and tender diffusely to deep palpation. Lower extremities had trace edema bilaterally. His neurologic exam was grossly non-focal, but asterixis was present.

Laboratory evaluation was remarkable for marked transaminitis, hyperbilirubinemia, elevated ammonia, and a moderate coagulopathy. In addition, he was pancytopenic and had pre-renal azotemia. A urine drugs of abuse screen was positive for opiates and cocaine. Andrew's parents consented for him to be tested for HIV, which later came back positive. A CT of the abdomen and pelvis demonstrated a large mass in the head of the pancreas, multiple liver nodules, and extensive intraabdominal lymphadenopathy – most consistent with a diagnosis of pancreatic cancer. Andrew's parents were notified of the findings to date, and the high likelihood that he had a terminal cancer. Subsequent tissue biopsy confirmed the diagnosis.

From: *Evidence-Based Medical Ethics*
By: J.E. Snyder and C.C. Gauthier © Humana Press, Totowa, NJ

The Ethical Dilemma

On admission, Andrew had been started on intravenous fluids and lactulose for presumed hepatic encephalopathy. With this intervention he began to show signs of slightly improved mental status and became increasingly more alert and conversive. However, he remained disoriented to time and date, and only intermittently knew he was "in the hospital." During a family meeting with the health care team, Andrew's parents requested that Andrew not be told of his pancreatic cancer diagnosis, even if he were to become more oriented. They argued that "he's had such a hard life" and that he should be spared the devastating news and just be made comfortable since his diagnosis was terminal. They also requested that his new HIV diagnosis be kept off his death certificate when he died. They stated that they came from a small town, and that "word gets out about such things" and they "don't want everyone in town knowing he had AIDS."

> **Questions for thought and discussion:** Can the medical team follow the request of Andrew's parents and not share the cancer diagnosis with him? What are the team's obligations to Andrew and his parents in this regard?

> **Questions for thought and discussion:** Under what circumstances, if any, should a health care provider withhold information from a patient? Is the assessment of a patient's capacity to make health care decisions also the determinant of their ability to handle information about their condition?

> **Question for thought and discussion:** Can the medical team agree to the other appeal of Andrew's parents, to not write HIV as a contributing factor on his death certificate when he passes away?

The Medicine

According to the CDC, the overall number of patients becoming newly infected with hepatitis C is declining, although 4.1 million Americans have been infected with the virus and 3.2 million are chronically infected. The most common route of transmission is through injection drug use (IDU). For persons with a five year history or more of IDU, it is estimated that 60 to 80 percent are infected with hepatitis C. Up to 10,000 people die annually due to hepatitis C-related liver disease, and annual health care costs related to hepatitis C are around $600 million.

Although male-to-male sexual contact remains the most common route of transmission in AIDS cases to date (59% of cases), the CDC reports that, as of 2005, approximately 22 percent of cases in adult and adolescent males are attributable to IDU alone, and an additional 8 percent is due to high risk heterosexual contact. In adult and adolescent females, IDU accounts for 40 percent of cases, whereas

56 percent are due to high risk heterosexual contact. About 30 percent of persons with five or more years' history of IDU will become infected with HIV.

The risk factors for developing pancreatic cancer are poorly understood, and there is no known correlation between the malignancy and hepatitis C or HIV infection. Pancreatic cancer is the tenth most common cancer among men, and ninth among women in the United States; however, it is the fifth most common cause of cancer death in the combined group. It is estimated that, in 2007, there will be 37,170 new cases of pancreatic cancer diagnosed, and 33,370 will die of the disease. The poor survival rates seen with pancreatic cancer are due largely to the fact that it is usually diagnosed late in its progression to a surgically-incurable, advanced stage, when symptoms first become prominent. Median survival after diagnosis is four to six months.

The Law

Until recently it was an accepted part of medical practice to withhold the truth about a terminal illness from the patient. Family members would often be informed about a dire diagnosis and prognosis, rather than informing the patient themselves. It was believed that knowing the truth would eliminate the possibility of hope in the patient, that they would give up and no longer fight for life, so that death might come sooner. This attitude began to change when the *Principle of Respect for Autonomy* came into prominence. Practitioners realized that patients would not be able to make plans for their last days, for their families, or for end-of-life care if they didn't know the truth about their medical conditions.

Although a number of court cases, such as *Canterbury v. Spence* (1972) and *Cobbs v. Grant* (1972), have established the legal right to information prior to consent for medical procedures, a legal right to the truth about a patient's medical condition has not been tested in the courts. Disclosure has usually been discussed in the context of voluntary informed consent, rather than disclosure of a patient's diagnosis and prognosis, when there are no viable treatment options.

According to the American Medical Association's Code of Medical Ethics, "Patients have a right to know their past and present medical status and to be free of any mistaken beliefs concerning their conditions." A new subsection, added in 2006, includes the following statement: "Withholding medical information from patients without their knowledge or consent is ethically unacceptable." The Ethics Manual of the American College of Physicians discusses the disclosure of uncomfortable information, as in cases where the "illness is very serious." Here, the manual recommends that "Upsetting news and information should be presented to the patient in a way that minimizes distress."

All states use a standard format for death certificates that is based on national recommendations from the National Center for Health Statistics. In Andrew's case, if the *primary cause of death* is not related to HIV infection, then it should not be listed in this section of the death certificate (Cause of Death, Part I). There is a second section on the certificate (Cause of Death, Part II), where other conditions that contribute to a patient's death – but did not result in the death itself – should be listed. It is at the certifying physician's discretion to list conditions that he or she finds to be relevant to the patient's death. One could argue that Andrew's HIV was not a contributing factor to his death and could potentially be excluded from the death certificate. As an official document of public record that will be used for the study of national health statistics and that the certifier seals with a signature, it is vital that the death certificate be completed with complete honesty and sound judgment, and that all pertinent information is documented.

The Ethics

In deciding whether or not to tell Andrew about his diagnosis, the medical team could use the *Principle of Beneficence* and the *Principle of Non-Maleficence*. Keeping this information from Andrew, as his parents have requested, may temporarily prevent the fear, anxiety, and other emotional suffering created by knowing that he is likely to die soon. However, these emotional reactions will occur later, as Andrew's cancer advances and his symptoms remain unexplained. While there is some immediate harm prevented, there is also harm done by withholding this information from Andrew.

If he doesn't know the truth, Andrew cannot make any decisions or plans for the end of his life, if he becomes more cognitively aware. Then he will need to decide if he wants to die at his parent's home, in the hospital, or in a hospice facility, and whether or not to have a Do Not Resuscitate Order. He is also likely to have financial decisions to make and even simple choices such as who to see and how to spend his last few months.

Even if Andrew is not able to make any of these decisions or be released from the hospital, he will be harmed later by the confusion, fear, and anxiety of not knowing why his condition is getting worse and he is being kept in the hospital.

Whatever Andrew's ability is to make health care decisions, it would be in his best interests to be told the truth about his illness. Most importantly, it is not up to his parents to make this decision. Andrew is the medical team's patient and it is Andrew's best interests that must determine their actions. Their responsibility, here, is to Andrew and not his parents, based on the patient-physician relationship.

According to the *Principle of Respect for Dignity*, the medical team should treat Andrew's parents with respect for their relationship with him and their emotions in this difficult situation. They naturally want to shield him from further suffering. However, respect for the parents' dignity is overridden, in this case, by respect for Andrew's autonomy and his dignity as a person. Andrew may recover his decision-making capacity so that he is able to make health care decisions for himself, for example to refuse or accept treatment, including experimental treatment for the pancreatic cancer, and whether or not to have CPR and other life-prolonging interventions at the end of life. Even if Andrew never regains the capacity to make serious medical decisions, he still needs to know he is dying so he can decide how and where he wants to spend his last days and make other simple life choices about his surroundings and his relationships with others.

According to the *Principle of Veracity*, Andrew must be told the truth about his medical condition while he is conscious, alert, and communicative, even if he is unable to make medical decisions or simple life choices. An important part of being respected as a person and being treated with respect and dignity is being told the truth about one's own medical situation. Andrew needs to be given the opportunity to understand what is wrong with him, the cause of his symptoms, and his ongoing medical care as the cancer advances.

There is no reason to withhold this kind of information from an adult patient who is conscious and communicating with the medical team. Andrew's ability to handle information about his condition is irrelevant. This is not something that the medical team can ever predict accurately and is not up to the team to determine. If patients do not understand the information provided to them, they will either ask questions or not, as their cognitive abilities permit. They must, at least, be given the chance to process this information, whether or not they can do so in the end.

Furthermore, the capacity to make medical decisions is not relevant to the patient's right to the truth. Even a young child or a patient with Alzheimer's disease deserves the respect, based on the *Principle of Respect for Dignity*, of being informed about what is being done to them and why. Although they may not be able to make decisions about their own treatment, they can be told, in language they can understand, why certain decisions are being made.

It seems clear that HIV should not be listed on Andrew's death certificate and would not be, even without the parents' request. If Andrew dies from the advancing pancreatic cancer, as seems likely, HIV will not have been the primary cause of death. It will not even have been a contributing factor. The only reason HIV might be put on the death certificate is if state law requires this for the protection of funeral home personnel or for the purpose of gathering vital statistics data. To adequately answer this question the medical team should ask the hospital legal department about relevant state laws concerning the identification of deceased HIV patients.

The Formulation

Now that the evidence-based medicine, legal precedent, and relevant ethical principles for this case have been reviewed, formulate a strategy to address the ethical conflicts in this case. If necessary, perform additional research into local and state laws and hospital regulations. Consider delving further into the background medical literature to assist with making sound therapeutic decisions. Devise a treatment approach that addresses the needs of the patient and his family, that is both ethically and medically sound, and that is culturally competent. Ensure that the strategy employs fair and appropriate utilization of medical resources, and that the approach is practical and feasible within the limits of the medical system at large. Work out a clear and professional way to communicate the proposal to the patient and his

family. Attempt to foresee challenges that may arise in conveying or implementing the plan. Determine what follow-up will be necessary to ensure that the chosen strategy remains successful for the patient in the long-term. Reflect on how the knowledge and skills learned from this case can be used to improve the care of patients that may be encountered in future practice.

Afterthoughts

In this case a family requested that certain information be kept from a patient to "spare him" the bad news. Would this case have been different if the patient were a child or an adolescent? What if the patient was elderly and the surrogate decision maker was their adult child?

Annotated References/Further Information

Code of Medical Ethics of the American Medical Association: Current Opinions with Annotations, 2006–2007 Edition. Council on Ethical and Judicial Affairs. Annotations prepared by the Southern Illinois University Schools of Medicine and Law. The subsections, "Patient Information" (8.12) and "Withholding Information from Patients" (8.082) are particularly relevant to this case.

Snyder L, JD, and Leffler C, JD. Ethics Manual, Fifth Edition. Ethics and Human Rights Committee, American College of Physicians. Annals of Internal Medicine. 142(7):560–582, 5 April 2005. The subsection, "Disclosure," contains useful guidance on providing distressing information to patients.

Canterbury v. Spence (1972). U.S. Court of Appeals, District of Columbia Circuit. 464 Federal Reporter, 2nd Series, 772.

Cobbs v. Grant (1972). California Supreme Court. 104 California Reporter 505.

http://www.cdc.gov/hiv/topics/surveillance/basic.htm#hivest. Accessed October 15, 2007.

National hepatitis C prevention strategy. A comprehensive strategy for the prevention and control of hepatitis C virus infection and its consequences. Summer 2001. Acquired from: *http://www.cdc.gov/ncidod/diseases/hepatitis/c/plan/index.htm*. Accessed October 15, 2007.

The National Association of Medical Examiners. *http://www.thename.org*. Accessed October 15, 2007.

U.S. Cancer Statistics Working Group. United States Cancer Statistics: 1999–2003 Incidence and Mortality Web-based Report. Atlanta: U.S. Department of Health and Human Services, Centers for Disease Control and Prevention and National Cancer Institute; 2007. Available at: *www.cdc.gov/uscs*. Accessed October 15, 2007.

Yendluri V et al. Pancreatic cancer presenting as a Sister Mary Joseph's nodule: case report and update of the literature. Pancreas. 34(1):161–4, 2007 Jan.

Van Cutsem E et al. Systemic treatment of pancreatic cancer. European Journal of Gastroenterology & Hepatology. 16(3):265–74, 2004 Mar.

Case 22
When a Patient Requires Significant Medical Resources

The Patient

Pamela B. is a 52-year-old female with a past medical history of alcohol and drug abuse, hepatitis C, cirrhosis, hepatic encephalopathy, and two prior episodes of upper gastrointestinal bleeding due to gastric ulcer. She presented to the Emergency Department with a two-week history of worsening ascites and a one-day history of hematemesis. Over the previous two weeks the patient noted worsening abdominal distention and, several days later, began to notice she was having black, tarry stools. On the morning of admission, she additionally had three episodes of large, frank hematemesis, prompting her to seek medical assistance. Pamela reported routine adherence with her home medications which included furosemide, spironolactone, and lactulose. She had a 30 pack-year history of tobacco use, prior use of heroin by injection, and ongoing consumption of up to a pint of hard liquor daily.

On physical examination, Pamela's vital signs were notable for a blood pressure of 92/57. She appeared to be uncomfortable and had mild jaundice. She was oriented to person and place, but not date or time. Her heart was regular and without murmurs, rubs, or gallops. Lungs were clear to auscultation bilaterally. Her abdomen had normoactive bowel sounds, but was grossly distended with ascites, and there was moderate diffuse tenderness to palpation. There was trace bilateral pedal edema. Neurological examination was normal. On rectal examination, external hemorrhoids were noted. Stool was black in color and strongly hemoccult positive.

A nasogastric tube was placed and red blood was aspirated. This did not clear with flushing, using 1L of normal saline. Laboratory evaluation was remarkable for a slightly elevated blood urea nitrogen (BUN) level, moderate transaminitis, a conjugated bilirubin of 4.1 mg/dL, and an international normalized ratio (INR) of 2.7. The Model for End-Stage Liver Disease (MELD) score was calculated to be 24. Her complete blood count was notable for a hemoglobin of 7.4 mg/dL, a hematocrit of 22.1 and a platelet count of 21,000. A urine drugs-of-abuse screen was negative and serum alcohol was not detected. Diagnostic paracentesis was consistent with ascites due to portal hypertension, and there was no evidence of peritonitis. A blood type and screen was sent, and a gastroenterology consult called for endoscopy. Serial hemoglobin/

From: *Evidence-Based Medical Ethics*
By: J.E. Snyder and C.C. Gauthier © Humana Press, Totowa, NJ

hematocrit levels were ordered for every four hours. The patient was started on intravenous fluids and a proton pump inhibitor, and was admitted to the hospital.

The Ethical Dilemma

Pamela's next hemoglobin and hematocrit resulted as 5.1 and 15.4, respectively, and so four units packed red blood cells were transfused, along with four units of fresh frozen plasma (FFP) and a 10-pack of platelets. Additionally 10 mg of vitamin K was administered. She was brought emergently to upper endoscopy where her stomach was noted to be full of blood, and an actively bleeding gastric ulcer was found. The area was cauterized, clipped in three places, and injected with epinephrine. Pamela was transferred to the intensive care unit where she was subsequently noted to have a continued drop in her hemoglobin/hematocrit. After more transfusions she was brought back to endoscopy and continued bleeding from her ulcer site was noted. More cautery, clipping, and epinephrine injections were done, but some oozing of blood from the ulcer site was still noted. The surgery team was consulted to consider partial gastrectomy, but Pamela was deemed too poor a surgical candidate due to a high intraoperative risk for mortality. Since reversal of coagulopathy did not occur with the previous vitamin K dose, transfusion with FFP was continued along with more red blood cells and platelets.

During her hospital course Pamela's mental status remained somewhat altered, presumably from hepatic encephalopathy. Her family was notified continuously of her poor condition. They agreed to a Do Not Resuscitate or Intubate (DNR/DNI) code status for her, but still wanted every other medical intervention done to try and save her life. A third upper endoscopy was deemed unlikely to be beneficial when her blood counts continued to trend down, despite having received a total of 20 units of packed red blood cells, 12 units of fresh frozen plasma, and six 10-packs of platelets. It was at this time that the hospital blood bank called the primary medical team with their concern about her high transfusion needs.

> **Questions for thought and discussion:** Should a limit be placed on how many blood products a patient should receive, based on resource availability? If so, how would one determine such a limit? If not, can one justify using resources up that may help other patients?

The Medicine

The MELD score is based on an equation that incorporates the serum bilirubin, creatinine, and INR, and has been a validated predictor of three-month mortality in end-stage liver disease (ESLD). The United Network for Organ Sharing (UNOS) uses MELD scoring to assist in allocating liver transplants to waiting list candidates, and helps prioritize recipients by need rather than wait-time. In a patient with a MELD score of 24, there is a predicted three-month mortality rate of 50 to 76 percent.

In 1996 Rockall, et al. developed a risk assessment system to evaluate prognosis in patients after upper gastrointestinal bleeding. Based on age, presence of shock, comorbid illness, etiology of the bleed, and stigmata of the hemorrhage, the risk of rebleeding and death may be estimated. Pamela has the highest possible score (eight) using the Rockall method, which is associated with a 41.1 percent mortality rate.

Massive blood transfusions may be required to treat certain patients, particularly those who have suffered trauma, gastrointestinal bleeding, and leaking abdominal aortic aneurysm. One study by Harvey, et al. looked retrospectively at the outcome of 43 such patients over a one year time period. Collectively, these patients consumed 16 percent of the medical center's blood products and had a 60 percent overall survival rate. Mortality was notably higher (74%) in those 19 patients who developed coagulopathy.

The 2005 Nationwide Blood Collection and Utilization Survey suggests that the nation's blood supply in the prior year was in a surplus status, with 648,000 more units of whole blood and packed red blood cells collected than transfused. There were 85.6 units of these specific products collected per thousand donor population, and 49.6 units transfused per thousand total population. Although the nation's overall transfusion needs were met over the year, shortages at local sites did occur. For example, 135 hospitals reported cancelling elective surgeries on a total 546 patients due to shortages in their blood product stores. In the survey the mean of the average amount paid per unit of red blood cells transfused was $201.07[*], per unit of fresh frozen plasma was $56.29[**], and per unit of whole blood-derived platelets was $63.67[***]. Using these mean per unit values, Pamela has already consumed $8,517.08 in blood products alone.

The Law

When Pamela presented to the Emergency Department, the medical team was correct to admit her to the hospital and attempt to control her bleeding with transfusions of blood products and endoscopy. Her need for blood was emergent and endoscopy was unsuccessful in controlling her bleeding, so the use of so many units of blood was appropriate at the beginning of her hospital course. However, once the hospital

---

[*] Assumes O-positive blood type, leukocyte-reduced, non-irradiated, not cytomegalovirus-negative

[**] Assumes type AB, with a 250mL volume

[***] Assumes not leukocyte-reduced, non-irradiated, and with three days remaining before product expires

blood bank called to express their concern about Pamela's high transfusion needs, the situation needed to be re-evaluated.

Under the Public Health Service Act (1975) hospitals receiving federal funds for building or renovation must provide emergency care to all patients residing in the service area, without discrimination on any basis unrelated to the patient's need or the availability of services. In this case, it would not violate federal law to limit the further use of the hospital's store of blood products for Pamela. Services must be provided if they are available, and Pamela's need for blood has begun to threaten the availability of blood products for other patients.

Any hospital in a mid-sized or large city must have blood products on hand for the emergency care of trauma victims, such as those in car accidents or those with gunshot or knife wounds. The hospital blood bank should use their records to determine how much blood they will need, until more becomes available. Then they must make sure they retain at least this amount, even if it means they cannot provide as much as may be needed by this patient.

Based on federal law, the medical team would have been legally protected had they informed Pamela's family that her blood transfusions would have to be limited based on availability and the needs of other patients. The medical team would also have been wise, in this case, to consult with the legal department before actually stopping Pamela's transfusions on the basis of availability, since she is very likely to die as a result.

The Ethics

Pamela's family is attempting to promote her best interests by asking that everything be done to save her life. The medical team, however, is in a better position to determine what would really be in Pamela's best interests, based on their ongoing monitoring of her condition. Employing the *Principle of Beneficence*, it would appear that continued blood transfusions would promote good for Pamela, in the sense of saving her life, and prevent the suffering and death that will result from her uncontrolled loss of blood. Since the attempts at surgical intervention have not been successful, continued blood transfusions seem to be the only way to acutely prevent her death.

However, turning to the *Principle of Non-Maleficence*, if the medical team cannot control the bleeding, they may be prolonging Pamela's suffering with continued blood transfusions. With her End-Stage Liver Disease, Pamela is likely to decline, even if her bleeding is controlled at this point. The medical team should consider her quality of life, as well as the length of her life. Considering the harm that may be done, it is not clear that it is in Pamela's best interests to continue the blood transfusions. The medical team may also be coming to the realization that the blood transfusions will have to be considered ineffective in controlling her bleeding.

The *Principle of Respect for Autonomy* will not apply directly to this case since Pamela is not able to express her wishes. However, the family can be engaged in a discussion about what they think she would want, given her condition and her chances for recovery. Thus, they will need to be fully and honestly informed about her condition and her prognosis, as the situation unfolds.

Based on the *Principle of Respect for Dignity* the family must be treated with respect for their emotions, their relationships with Pamela, and their reasonable goals for her. If they do not know what she would want, they should consider what would be in her best interests in the long-term. They should be encouraged to consider her quality of life, now that attempts at controlling her bleeding have proven to be unsuccessful. They may also need to know she has End-Stage Liver Disease and is not going to be acutely a candidate for a liver transplant since she is continuing to have active bleeding and has a history of ongoing alcohol abuse. This may help them realize that their original goals for her treatment are no longer realistic.

At some point the medical team may determine that continued transfusions would be futile or ineffective in meeting any reasonable goals for Pamela. The family needs to know this and be provided with the reasons for this judgment, in language they can understand.

This case raises questions about the distribution of health care resources. The *Principle of Distributive Justice* requires that health care resources be distributed in a fair way among the members of society. Although "fairness" is a particularly vague concept, one attempt by the federal government to insure fairness was the Public Health Service Act which states that all patients deserve emergency treatment as long as the care they need is readily available. Other justice-based federal programs include Medicaid and Medicare.

The question in this case is: would using all of the hospital's blood products for one patient be a fair distribution of this medical resource? The medical team should not make this determination. As the Ethics Manual of the American College of Physicians recommends, "Resource allocation decisions are most appropriately made at the policy level rather than entirely in the context of an individual patient-physician encounter." The decision about if and when to limit transfusions for Pamela should be made by hospital administrators, in consultation with the blood bank personnel.

However, the medical team may bring the concerns of the blood bank to the attention of Pamela's family. The family members may want to have this information as they continue to make decisions about her care. The fact that Pamela's transfusions are not proving to be effective should be discussed, along with the fact that the supply of blood products for other patients is becoming dangerously low.

Notice, too, that if the blood transfusions had been successful in halting Pamela's blood loss, her ability to pay for the blood should be irrelevant. It should also not be a question of whether or not Pamela *deserves* these resources, based on her contribution to society or some measure of her merit as a person.

The hospital administrators should consider two factors in making this decision. First, they should consider the effectiveness of the relevant medical resource, based on Pamela's medical condition and prognosis. Can her bleeding be controlled? Is her need for blood products expected to end? Is she expected to recover? Second, they should consider the availability of the resource in question. Do they now have enough blood products to cover the normally expected need by other patients in the time period before they are able to obtain more? If the decision is made to limit further blood transfusions for Pamela, the family needs to be given an explanation, in terms of their effectiveness for her and the need to retain enough blood to meet the needs of other patients.

The Formulation

Now that the evidence-based medicine, legal precedent, and relevant ethical principles for this case have been reviewed, formulate a strategy to address the ethical conflicts in this case. If necessary, perform additional research into local and state laws and hospital regulations. Consider delving further into the background medical literature to assist with making sound therapeutic decisions. Devise a treatment approach that addresses the needs of the patient and her family, that is both ethically and medically sound, and that is culturally competent. Ensure that the strategy employs fair and appropriate utilization of medical resources, and that the approach is practical and feasible within the limits of the medical system at large. Work out a clear and professional way to communicate the proposal to the patient and her family. Attempt to foresee challenges that may arise in conveying or implementing the plan. Determine what follow-up will be necessary to ensure that the chosen strategy remains successful for the patient in the long-term. Reflect on how the knowledge and skills learned from this case can be used to improve the care of patients that may be encountered in future practice.

Afterthoughts

In this case a patient needed large volumes of blood products to be kept alive, but there is a limited supply of such products and the question of appropriate resource utilization arose. Are there other situations, aside from blood products, where care may be limited by available resources? What are they?

Can a patient's case ever be considered futile based on resource availability? On cost? On the burden of care? How does one balance the esoteric concept of the value of a human life with the cost to the public?

Annotated References/Further Information

Code of Medical Ethics of the American Medical Association: Current Opinions with Annotations, 2006–2007 Edition. Council on Ethical and Judicial Affairs. Annotations prepared by the Southern Illinois University Schools of Medicine and Law.

Snyder L, JD, and Leffler C, JD. Ethics Manual, Fifth Edition. Ethics and Human Rights Committee, American College of Physicians. Annals of Internal Medicine. 142(7):560–582, 5 April 2005. The subsection on Resource Allocation is particularly relevant to this case.

Amico, GE et al. Natural history and prognostic indicators of survival in cirrhosis: a systematic review of 118 studies. Journal of Hepatology. 44(2006):217–231.

Kamath PS et al. A model to predict survival in patients with end-stage liver disease. Hepatology. 33(2):464–470. 2001.

Malinchoc M et al. A model to predict poor survival in patients undergoing transjugular intrahepatic portosystemic shunts. Hepatology. 31(4):871. 2000.

http://www.unos.org/resources/glossary.asp#M. Accessed October 15, 2007.

The Department of Health and Human Services and AABB 2005 National Blood Collection and Utilization Report, acquired from:

http://www.aabb.org/apps/docs/05nbcusrpt.pdf. Accessed October 15, 2007.

http://www.bloodcenters.org/aboutblood/questions.htm#footnote. Accessed October 15, 2007.

Rockall TA et al. Risk assessment after acute upper gastrointestinal hemorrhage. Gut. 1996;38: 316–21.

Harvey MP et al. Massive blood transfusion in a tertiary referral hospital. Clinical outcomes and haemostatic complications. Medical Journal of Australia. 163(7):356–9, 1995 Oct 2.

Public Health Service Act (1975) Title VI. 42 Code of Federal Regulations Part 124.

www.hhs.gov/ocr/hburton.htm. Accessed October 15, 2007.

Case 23
When a Patient's Belief System Affects His Care

The Patient

Brian J. is a 42-year-old male without significant prior medical history who presented to the Emergency Department with acute onset of right upper quadrant pain several hours after consuming a meal at a fast food restaurant. He described the pain as sharp in nature, with radiation to his right shoulder. He also complained of nausea, with two episodes of non-bloody emesis. He had no recent change in bowel function, and denied hematochezia or melena. He denied fevers, but had some chills and diaphoresis. He had no known drug allergies and took no medications at home. He denied alcohol, tobacco, or illicit drug use. He was married with one child, college-educated, and worked in an office performing computer technical support. Family history was noncontributory.

On physical exam, there was a low-grade fever and stable vital signs. He was anicteric and mucous membranes were moist. Heart and lung exam were unremarkable. His abdomen had slightly hypoactive bowel sounds, was non-distended, and was markedly tender in the right upper quadrant with a positive Murphy's sign. Rectal exam was normal with negative hemoccult testing. Extremities were warm and well-perfused, without edema. Neurological exam was grossly non-focal.

Laboratory evaluation was notable for a mild leukocytosis, transaminitis, and elevated alkaline phosphatase. Chest radiograph and electrocardiogram were normal. A right upper quadrant sonogram showed the presence of multiple gallstones, gallbladder wall thickening, and surrounding edema, all consistent with a diagnosis of acute cholecystitis. Brian was admitted to the general medical-surgical floor, made *nil per os*, and started on intravenous fluids and antibiotics. A surgical consult was called to evaluate for cholecystectomy.

The Ethical Dilemma

When the surgical resident was consenting Brian for cholecystectomy and described the risks and benefits of the procedure, one of the risks was the potential for bleeding. Brian told the resident that, as a Jehovah's Witness, he could not accept a blood transfusion

From: *Evidence-Based Medical Ethics*
By: J.E. Snyder and C.C. Gauthier © Humana Press, Totowa, NJ

were bleeding to occur. The resident stated to Brian that she was unsure if the surgical team could proceed with the surgery if the possibility of emergency transfusion were ruled out, and that she needed to discuss the matter further with her attending.

Questions for thought and discussion: What are the beliefs of Jehovah's Witnesses regarding medical care and blood transfusion? Are any blood products ever allowable? Are blood substitutes acceptable?

Questions for thought and discussion: If a Jehovah's Witness will not accept a blood transfusion in the case of significant bleeding, should an elective procedure be withheld from them? Should an emergent procedure be withheld?

Questions for thought and discussion: In general, how much risk is acceptable to move forward with a procedure on a patient? How does one balance respect for a patient's belief system and yet offer the patient the best possible care?

The Medicine

About three-quarters of patients with acute cholecystitis will have their symptoms resolve within one week of initiating medical management alone. The others will require immediate surgical intervention due to development of a complication such as empyema, gangrene, or perforation. Of those managed initially without surgery, a majority (60%) are likely to have a recurrence of symptoms in the coming years. As a result, eventual surgical treatment of all patients who develop acute cholecystitis is generally considered to be the standard of care for those healthy enough to undergo the operation. Laparoscopic cholecystectomy has become the most popular surgical procedure to treat gallbladder stones in the United States since it was developed in 1988. Lee, et al. reviewed six papers that investigated complication rates from the procedure and noted that the risk of "major bleeding" from the procedure varied from 0.2 to 4.3 percent. Most major bleeds are due to injuries of the epigastric vessels, cystic artery, or the liver bed, and may result in conversion to an open procedure. Overall mortality for laparoscopic cholecystectomy is low, estimated between 0 and 0.13 percent.

Although they have no objection to medical care in general, or to surgery, Jehovah's Witnesses are not permitted to accept allogeneic or autologous (pre-stored) transfusion of whole blood, packed red blood cells, plasma, white blood cells, or platelets during surgery or otherwise. This is based on their deeply held belief that the Bible prohibits the "ingestion" of blood as cited in several Bible verses, such as "You must not eat blood" (Leviticus 7:26) and "Keep abstaining from blood" (Acts 15:29). The rules about accepting blood products are complex, however, particularly around topics such as cell salvage and blood cell tagging. Individual Witnesses should be asked what their wishes are regarding the use of blood products or fractions thereof. The official Jehovah's Witness website is a helpful resource for providers, and patients may want to consult with church leaders for advice.

An area of growing interest for Jehovah's Witnesses, as well as for others who do not want to receive transfused blood products, is so-called *bloodless surgery*. This often involves the administration of iron and/or erythropoietin preoperatively to bolster a patient's red blood cell reserves, the use of minimally invasive techniques and low central venous pressure anesthesia, the ligation or electrocautery of bleeding vessels intraoperatively, minimizing postoperative phlebotomy, and the use of non-blood volume expanders in the setting of shock. Some Witnesses will allow use of the Cell Saver, whereby autologous blood is salvaged during a procedure.

Blood substitutes not only offer another potential alternative to blood transfusion for Jehovah's Witnesses, but could also be a viable option for others who do not wish to receive banked blood. In addition, such products do not depend on blood bank supplies, significantly lower the risk of transfusing infectious agents, and virtually eliminate immune-modulated transfusion reactions. These products may function both as volume expanders and oxygen carriers. Hemoglobin-based solutions and perfluorocarbons are being researched as potential substitutes for red blood cell products, and are currently in various stages of development and trials.

The Law

Case law in states throughout the country has established the legal right of adult Jehovah's Witnesses to refuse blood transfusions based on their religious beliefs, even if the refusal is likely to result in death. The case of *Public Health Trust v. Wons* (1989) concerned a Witness who refused a blood transfusion for uterine bleeding. The hospital was permitted by the circuit court to give the transfusion, against her will, because she had two minor children. However, the Florida Court of Appeals and the Florida Supreme Court both ruled that the state's interest in preserving life and providing a home with two parents for the minor children did not override the patient's constitutional rights to practice her religion and to privacy.

The Patient Self-Determination Act (1990) has also strengthened the right of Jehovah's Witnesses to refuse blood products. This federal law requires health care institutions receiving Medicaid and Medicare funds to provide written information to all adult patients concerning their rights to make decisions about their medical care, "including the right to accept or refuse medical or surgical treatment." With this law in place, there is little doubt that a competent adult Witness must be permitted to refuse blood transfusions.

The Ethics

The surgical team will need to consider the risks and benefits of doing this surgery without the possibility of an emergency blood transfusion. Using the *Principle of Beneficence* they will consider the good promoted, such as a return to good health for Brian, and the harm removed and prevented, including severe pain, infection, and the possible perforation of the gallbladder. Of particular importance, in this case, is the *Principle of Non-Maleficence* which requires the practitioners to avoid causing harm to their patients. Attempting this surgery without the option of a blood transfusion may risk having uncontrolled bleeding and possibly result in death. Comparing these risks and benefits, the surgical team will need to determine what would be in Brian's best interests.

The surgical team may decide to attempt the surgery, even under this restriction, given the low rate of a major bleed and the low mortality rate. There are two things they could do, however, to minimize these risks. First, they could talk with Brian about his willingness to accept components such as albumin and blood substitutes. They should also let him know that, if a bleed occurs, they may need to convert to an open procedure to enable use of methods other than a transfusion to control the bleeding. They should ask for his consent to change the procedure, in case this becomes necessary.

The *Principle of Respect for Autonomy* is particularly important in this case as the basis for the Patient Self-Determination Act and the court rulings that permit Jehovah's Witnesses to refuse blood transfusions. Brian has agreed to the recommended surgery, but has refused one aspect of the standard of care for that particular surgery. According to this principle, Brian must be allowed to refuse the transfusion of blood products.

However, the principle does not mean that the surgical team is required to go ahead with the surgery under this condition. This principle only permits capable patients to accept or refuse recommended treatment. If the surgical team decides not to attempt the surgery without the option of a blood transfusion, they will need to notify Brian that their recommendation is that he have the surgery with the possibility of a transfusion or not have it at all.

An essential part of respecting the autonomy of capable patients is the requirement of voluntary informed consent. Brian cannot make an informed decision to undergo the surgery without a blood transfusion, if needed, unless he is fully informed and understands the risks and alternatives as well as the expected outcome of not having the surgery at all. If the surgical team decides to go ahead with the surgery, they need to be careful and thorough in informing Brian of the increased risks and the possible alternatives for controlling a major bleed. It will also be important for Brian to sign a waiver that indicates his understanding of these risks and his willingness to undertake them.

The other essential principle in this case is the *Principle of Respect for Dignity*. Respect for Brian's dignity as a person requires respect for his religious beliefs and his deep commitment to them. In the past physicians have sometimes been judgmental about the religious beliefs of Jehovah's Witnesses and refused to proceed with needed surgery because of the limitations their beliefs placed on the established standard of care. This attitude is changing and is encouraged by the new Bloodless Medicine and Surgery Programs at many hospitals. These programs educate practitioners and support patients and practitioners as they develop treatment plans that meet the spiritual requirements of Jehovah's

Witnesses. Often a nurse administrator will coordinate the program, work with elders of the church in exploring alternatives to the transfusion of blood products, and supervise nurses caring for Jehovah's Witness patients ("The Pregnant Jehovah's Witness," 2005).

Early communication and planning for the care of Jehovah's Witnesses is essential. Even in the case of emergency surgeries, hospital administrators can be better prepared by meeting with the elders in their community to develop a program for providing respectful care to Witnesses in the area. These plans should include education for practitioners and a list of acceptable blood substitutes for various medical conditions and surgeries.

With such plans in place, it should not be necessary to refuse to provide either elective or emergency procedures to Jehovah's Witnesses. Even without such programs or plans, practitioners should approach their Witness patients with respect for their religious beliefs. They will need to consider the risks and benefits of the procedures their patients need, with the restriction against blood transfusions, and try to provide the best care they can within these restrictions. They should also explore the possibility of blood substitutes with their patients to reduce these risks.

In general, if the risks of the procedure without the option of providing blood products are equal to or less than the risks of forgoing the procedure, practitioners should proceed. The most important requirement, when risks need to be balanced, is to make sure the patient is fully informed of the risks that are being compared. Patients who are Witnesses may also choose to take greater risks with procedures performed without blood products than those they would face by forgoing these procedures altogether. In this case, practitioners need to decide for themselves if they want to proceed with these more risky procedures.

The Formulation

Now that the evidence-based medicine, legal precedent, and relevant ethical principles for this case have been reviewed, formulate a strategy to address the ethical conflicts in this case. If necessary, perform additional research into local and state laws and hospital regulations. Consider delving further into the background medical literature to assist with making sound therapeutic decisions. Devise a treatment approach that addresses the needs of the patient and his family, that is both ethically and medically sound, and that is culturally competent. Ensure that the strategy employs fair and appropriate utilization of medical resources, and that the approach is practical and feasible within the limits of the medical system at large. Work out a clear and professional way to communicate the proposal to the patient and his family. Attempt to foresee challenges that may arise in conveying or implementing the plan. Determine what follow-up will be necessary to ensure that the chosen strategy remains successful for the patient in the long-term. Reflect on how the knowledge and skills learned from this case can be used to improve the care of patients that may be encountered in future practice.

Afterthoughts

In this case a question arose about offering a procedure to a Jehovah's Witness patient if a complication such as bleeding may occur and blood transfusion was not an option. Are there any other situations in medicine where a patient's beliefs may interfere with what is considered standard of care? Are there any situations where a provider's belief systems affects their ability to deliver quality medical care to their patient?

Annotated References/Further Information

Code of Medical Ethics of the American Medical Association: Current Opinions with Annotations, 2006–2007 Edition. Council on Ethical and Judicial Affairs. Annotations prepared by the Southern Illinois University Schools of Medicine and Law.

Snyder L, JD, and Leffler C, JD. Ethics Manual, Fifth Edition. Ethics and Human Rights Committee, American College of Physicians. Annals of Internal Medicine. 142(7):560–582, 5 April 2005.

Acute and Chronic Cholecystitis. Harrison's Internal Medicine, Chapter 305. Acquired online from McGraw-Hill's Access Medicine on October 15, 2007.

Lee VS et al. Complications of laparoscopic cholecystectomy. The American Journal of Surgery. 165:527–532. April 1993.

Magner D et al. Pancreaticoduodenectomy after neoadjuvant therapy in a Jehovah's witness with locally advanced pancreatic cancer: case report and approach to avoid transfusion. The American Surgeon. 72(5):435–437. May 2006.

Watchtower – The Official Web Site of Jehovah's Witnesses. *http://www.watchtower.org/*. Accessed October 15, 2007.

Patient Self Determination Act (1990). 42 U.S.C. 1395 cc (a). Subpart E. Section 4751.

The Patient Self-Determination Act can be found at *www.dgcenter.org/acp/pdf/psda.pdf*. Accessed October 15, 2007.

Public Health Trust v. Wons (1989). Florida Supreme Court. 541 So. 2d 96.

Towarelli T and Valenti J. The pregnant Jehovah's Witness: how nurse executives can assist staff in providing culturally competent care. JONA'S Health Care Law, Ethics, and Regulation, Vol. 7, No. 4, October-December, 2005, pp. 105–109.

www.jehovah.to/. Accessed October 15, 2007. This website includes links to a number of Jehovah's Witness blood transfusion cases.

Jahr JS et al. Blood substitutes and oxygen therapeutics: an overview and current status. American Journal of Therapeutics. 9(5):437–43, 2002 Sep-Oct.

Winslow RM. Blood substitutes. Current Opinion in Hematology. 9(2):146–51, 2002 Mar.

Case 24
When a Minor Requires Confidential Medical Care

The Patient

Amy A. is a 15-year-old female without significant prior medical history who presented to the Emergency Department with complaints of mild pelvic pain, dysuria, and vaginal discharge for several days. She denied fevers or other constitutional symptoms. There were no joint complaints. Menstrual periods had been normal, with menarche at age nine and last menstrual period was four weeks previously. She reported no prior sexual activity, took no medications, and had no known drug allergies. She denied ever using alcohol, tobacco, or drugs. She was currently a high school sophomore and reported good school performance. Family history was noncontributory.

On physical exam Amy was afebrile and vital signs were stable. Head and neck, cardiovascular, and lung examinations were normal. There was mild suprapubic tenderness to palpation of the abdomen. Extremities were unremarkable and neurological exam was grossly non-focal. Pelvic exam was notable for the absence of cervical motion tenderness and the presence of yellowish discharge from a mildly erythematous cervix. A sample of the fluid was collected for testing for gonococcus and Chlamydia by DNA probe, and the gonococcus resulted as positive. Laboratory evaluation was also notable for a negative urine toxicology screen and a positive urine pregnancy test.

The Ethical Dilemma

When the results of the evaluation were presented to Amy she began to cry. She stated that she had a primary care doctor, but came to the Emergency Department because she was afraid she might have a sexually transmitted disease and didn't want that information on her medical chart. She did not expect the positive pregnancy test findings. Amy confessed that she had become sexually active with one male partner over the past several months, that he was 16-years-old, and that the intercourse was consensual. She implored that "none of this" can be told to her parents until she determines "what to do." She stated

From: *Evidence-Based Medical Ethics*
By: J.E. Snyder and C.C. Gauthier © Humana Press, Totowa, NJ

that she was considering an abortion as one of her potential options, but needed time to think about this further.

> **Questions for thought and discussion:** What are the the treating physician's obligations to Amy with regard to confidentiality, given that she is a minor? Is there an obligation to share any of the gathered information with Amy's parents? Are any of her conditions "reportable?"

> **Questions for thought and discussion:** Can Amy consent to receive medical treatment without her parent's knowledge or approval? Can she consent to HIV testing? Drug testing? Elective procedures? Emergency surgeries?

The Medicine

The CDC estimates that $14.1 billion dollars are spent annually in the direct treatment of sexually transmitted infections (STIs), and that almost half of the 19 million new STIs diagnosed annually occur in persons age 15 to 24. After Chlamydial infection, gonorrhea is the second most commonly diagnosed STI, with a total of 339,593 cases diagnosed in 2005. This may be a significant underestimate, given a likely high rate of underreporting. The prevalence of gonorrhea and other STIs is significantly higher in ethnic minority populations, likely due in part to limited access to medical care. Although gonorrhea is highly treatable with antibiotics, untreated it can result in pelvic inflammatory disease, ectopic pregnancy, infertility, Fitz-Hugh-Curtis syndrome, infectious arthritis, and disseminated infection.

According to UNICEF data, up to 63 percent of girls under age 18 report having had sexual intercourse, with 81 percent reporting having had sex by age 20. The United States has the highest teenage birthrate in the developed world, at 52.1 births per 1,000 girls aged 15- to 19-years-old. It is estimated that 22 percent of 20-year-old American women have had a child as a teenager. It is important to note, however, that the overall rate of births to teenagers is declining in the United States. This trend is attributed by some groups to more effective use of contraception and not increased rates of abstinence.

The rate of elective abortions to women under age 20 is 30.2 per 1,000. Legally performed abortions are considered less dangerous (less than one death per 100,000 events) than childbirth (nine deaths per 100,000 events). The advent of so-called "emergency contraception" or the "morning after pill" has no doubt also influenced the decline in the birth rate among teenage girls. Although over-the-counter in some European countries, it is available only by prescription in the United States. The treatment consists of taking two high-dose oral contraceptive pills, 12 hours apart, within 72 hours of the unprotected intercourse. Its purpose is to prevent ovulation, fertilization, or implantation. Taken as directed, it has up to a 95 percent likelihood of preventing pregnancy.

Although medically there is no increased risk for a teenager to carry a child to term than a woman in her 20s and beyond, data suggest that there potentially are significant psychosocial disadvantages to both the teenage mother and child. The mother is more likely to not finish her education, to be unemployed, to receive welfare, to have poor quality housing, and to be diagnosed with depression. The child is more likely to grow up fatherless, poor, to suffer from abuse or neglect, to have poor school performance, to abuse alcohol and drugs, and to become a teenage parent themselves.

The Law

Laws mandating the reporting of certain communicable diseases, such as gonococcal infection, have been covered previously (see Table 12.2). The age of the patient does not affect the obligation of a health care provider to report these conditions to the regional public health department.

The landmark, controversial, and politically charged 1973 U.S. Supreme Court case of *Roe v. Wade* (410 U.S. 113) was based on the legal challenge of a single, pregnant woman (dubbed "Jane Roe") against the state of Texas, represented by the District Attorney (Henry Wade) of Dallas County. In brief, the disputed Texas statutes held that an attempt at abortion for reasons other than to save the mother's life was a criminal act. The Supreme Court declared, by a 7 to 2 majority vote, that such a state law violates a woman's right to privacy, as protected in the 14th Amendment of the U.S. Constitution ("nor shall any State deprive any person of life, liberty, or property, without due process of law"). Hence, all state and federal laws conflicting with this decision were consequently voided. The ruling allowed women to elect abortion for any reason until a fetus is "viable" to survive on its own, including by artificial means. The usual time of viability was declared to lie between 24 and 28 weeks of gestation. Attempts to overturn the Supreme Court decision have not been successful to date. Interestingly, "Jane Roe" is now known to be Norma McCorvey, a Dallas woman who initially sought an abortion after claiming that rape was the cause of her pregnancy. Once a staunch abortion rights advocate, McCorvey has since rescinded her claim of being raped. She has also founded a ministry called "Roe No More," a pro-life advocacy group dedicated to fighting abortion rights.

In most states minors may receive a variety of medical services without the knowledge or consent of their parents. These include contraceptive and pregnancy services, as well as treatment for sexually transmitted and other reportable diseases, drug and alcohol abuse, and emotional disturbances. In these cases the confidentiality of minors is respected and the treating physician is legally obligated not to notify the minor's parents or discuss these services or treatments with them, without the minor patient's permission. However, sexually transmitted infections, such as gonorrhea, are considered "reportable" diseases and must be reported to the public health department.

Along with protection for the minor patient's confidentiality regarding these services, the minor is legally permitted to give consent for them on her own behalf. For example, she may give consent for HIV testing as well as drug testing. In the case of emergency procedures, if the minor is able to give consent, she should be allowed to do so. However, physicians may also initiate emergency procedures when a minor is not able to give consent and the parents cannot be contacted for permission in a timely manner.

Elective surgeries are more difficult. It is unlikely that elective surgeries will be related to any of the services and treatments for which a minor is legally able to give consent, except for pregnancy services. Therefore, it is strongly suggested that parents be involved in making decisions with their minor child about elective surgeries. Currently, in most states, a minor must first obtain the consent of her custodial parent, grandparent, or legal guardian prior to having an abortion. There is, however, the option of a judicial bypass in which the minor goes before a judge to explain the reasons why she cannot obtain the required consent.

The American Medical Association Code of Medical Ethics recognizes the legal right of minors to give consent for these services and treatments, and to have their confidentiality respected. However, it is also recommended that physicians encourage minors to discuss these medical issues with their parents and explore their reasons for not involving their parents.

Physicians should educate themselves about the applicable laws in the states where they practice medicine. Practitioners could contact the legal department of their hospital, for example, for guidance on these laws and relevant hospital policies.

The Ethics

Consent and confidentiality for minors, in certain sensitive areas of physical and mental health, are supported by the *Principles of Beneficence* and *Non-Maleficence*. When minors are allowed to give consent for these services, knowing their confidentiality will be protected, they are much more likely to seek the tests and treatments they need to preserve their health and to prevent the harm caused by untreated medical conditions. For example, Amy came to the Emergency Department because she thought her confidentiality would be better protected there than by going to see her primary care physician. Because she was able to give consent for sexually transmitted infection testing, she was diagnosed and will now receive treatment, preventing the potential effects of untreated gonorrhea, such as pelvic inflammatory disease and infertility.

On the other hand, violating the confidentiality of minor patients who are seeking these health care services could cause harm to the patient. Notifying the minor's parents of her tests results, diagnoses, and treatments may cause loss of trust in medical practitioners, thereby discouraging Amy from seeking health care in the future. Violating confidentiality could also damage family relationships beyond repair.

According to the *Principle of Respect for Autonomy* capable patients should be allowed to make their own health care decisions, including accepting or refusing recommended procedures. Amy has asked that her parents not be informed of her gonorrhea or her pregnancy. Amy, even at 15, may be capable of making her own medical decisions. One positive indication is that she was responsible enough to come to the Emergency Department for medical treatment. Morally, Amy should be permitted to make decisions about the procedures recommended by the treating physician, if that physician believes she is capable of doing so.

If she is judged to be capable, the treating physician should follow the guidelines for voluntary informed consent. Amy should be fully informed about the risks and benefits of these procedures, she must understand this information and her consent should not be the result of any coercion or undue pressure. She also needs to know that her gonorrhea must be reported to the public health department and the reasons for this should be explained to her.

The *Principle of Respect for Dignity* is particularly relevant to this case. Even minor patients deserve respect for their dignity, including respect for their emotions, relationships, reasonable goals, privacy, and bodily integrity. Allowing Amy to give consent for medical services shows respect both for her reasonable goals and her bodily integrity. She knew something was wrong and she came to the Emergency Department with the reasonable goal of seeking medical care. Allowing her to give consent for these tests and treatments gives her some control over what happens to her body.

Respecting Amy's confidentiality, in regard to her parents, expresses respect for her emotions, her relationships, and her privacy. Amy has greater knowledge about her relationship with her parents than the treating physician, and her emotional reaction to the news of her gonorrhea and pregnancy are based, in part, on that relationship. Her emotions and her privacy must be respected as essential aspects of her dignity.

In discussing her pregnancy with Amy, the treating physician should inquire about Amy's reasons for not involving her parents in her decision and ask whether she would like to receive the confidential services of the hospital's social worker or counselor, if available, to discuss her options. As an alternative, she should be encouraged to talk with another adult relative, friend, teacher, or minister. She should also be told that, depending on state laws, she will need parental consent if she chooses abortion, with the option of a judicial bypass if there are serious reasons why she can't discuss her pregnancy and choice of abortion with her parents.

The Formulation

Now that the evidence-based medicine, legal precedent, and relevant ethical principles for this case have been reviewed, formulate a strategy to address the ethical conflicts in this case. If necessary, perform additional research into local and state laws and hospital regulations. Consider delving further into the background medical literature to assist with making sound therapeutic decisions. Devise a treatment approach that addresses the needs of the patient and her family, that is both ethically and medically sound, and that is culturally competent. Ensure that the strategy employs fair and appropriate utilization of medical resources, and that the approach is practical and feasible within the limits of the medical system at large. Work out a clear and professional way to communicate the proposal to the patient and, if necessary, her family. Attempt to foresee challenges that may arise in conveying or implementing the plan. Determine what follow-up will be necessary to ensure that the chosen strategy remains successful for the patient in the long-term. Reflect on how the knowledge and skills learned from this case can be used to improve the care of patients that may be encountered in future practice.

Afterthoughts

In this case a minor sought medical treatment for symptoms that revealed a pregnancy and a sexually transmitted infection. Amy said she was considering an elective abortion and declined to involve her parents in her care. How would this case have been different if the patient was found to be abusing illicit drugs? If she was not pregnant, but was using illicit drugs, should this information be shared with her parents? How would this case have been different if the patient were a victim of sexual assault, statutory rape, or incest?

Health care practitioners come from diverse cultural and religious backgrounds, and may have preconceived notions about teenage pregnancy and abortion. How can a health care provider's own personal views affect their ability to offer a patient care? How strongly should a patient, minor or otherwise, be encouraged by their health care provider to seek alternatives to abortion, such as adoption?

Annotated References/Further Information

Code of Medical Ethics of the American Medical Association: Current Opinions with Annotations, 2006–2007 Edition. Council on Ethical and Judicial Affairs. Annotations prepared by the Southern Illinois University Schools of Medicine and Law. The subsections on "Confidential Care for Minors" (5.055) and "Mandatory Parental Consent to Abortion" (2.015) are recommended for practitioners who are treating minors.

Snyder L, JD, and Leffler C, JD. Ethics Manual, Fifth Edition. Ethics and Human Rights Committee, American College of Physicians. Annals of Internal Medicine. 142(7):560–582, 5 April 2005.

Weinstock H et al. Sexually transmitted diseases among American youth: incidence and prevalence estimates, 2000. Perspectives on Sexual and Reproductive Health. 2004;36 (1):6–10.

Trends in reportable sexually transmitted diseases in the United States, 2005. National surveillance data for chlamydia, gonorrhea, and syphilis. Centers for Disease Control and Prevention (CDC). Acquired from: *http://www.cdc.gov/std/stats/trends2005.htm#trendsgonorrhea*. Accessed October 15, 2007.

UNICEF, A league table of teenage births in rich nations. Innocenti Report Card No.3, July 2001. UNICEF Innocenti Research Centre, Florence.

Grimes DA. The morbidity and mortality of pregnancy: still risky business. American Journal of Obstetrics & Gynecology. 170(5 Pt 2):1489–94, 1994 May.

U.S. Supreme Court, Roe v. Wade, 410 U.S. 113 (1973). Acquired from: *http://caselaw.lp.findlaw. com/scripts/getcase.pl?navby=CASE&court=US&vol=410&page=113.* Accessed October 15, 2007.

The U.S. Constitution Online. Amendment 14. Acquired from: *http://www.usconstitution.net/ xconst_Am14.html.* Accessed October 15, 2007.

Wood, DS. Who is 'Jane Roe'? Anonymous no more, Norma McCorvey no longer supports abortion rights. CNN. Wednesday, June 18, 2003. Posted: 8:56 AM EDT (1256 GMT). Acquired from: *http://www.cnn.com/2003/LAW/01/21/mccorvey.interview/.* Accessed October 15, 2007.

Case 25
When a Patient's Condition Is Terminal

The Patient

Betty R. is a 75-year-old female with a past medical history of unresectable adenocarcinoma of the pancreatic head, with known hepatic metastases, that was diagnosed seven months prior to the current admission. She had previously undergone a trial of palliative chemotherapy with minimal relief of her symptoms of abdominal pain, and subsequently had a celiac nerve block with modest results. Two months prior to admission she underwent endoscopic biliary stent placement to relieve malignant obstruction, and tolerated the procedure well. She had since been living with family at home, with the assistance of outpatient hospice services. She presented to the Emergency Department for the current admission with complaints of worsening anorexia, weight loss, and jaundice over the preceding two weeks, and recurrence of severe abdominal pain. Her pain had been incompletely relieved with opiate medications at home, and instead of contacting the hospice nurse per usual protocol, her family brought her to the Emergency Department out of desperation to quickly relieve her symptoms. On presentation, limited history could be obtained from Betty herself, who was lethargic and visibly uncomfortable.

Betty's other past medical history was significant only for hypertension. Home medications included a thiazide diuretic, opiate analgesics, anti-emetics, and megestrol acetate to stimulate appetite. She had no known drug allergies. Family history was noncontributory. She had no history of tobacco, alcohol, or illicit drug use.

On physical exam Betty responded to voice, but was minimally verbal. She was notably jaundiced and uncomfortable, often moaning. Mucous membranes were very dry. She was tachycardic and her lungs were clear to auscultation. Her abdomen was diffusely tender even with gentle palpation, with positive guarding and rebound. The left lower extremity was swollen and tender in the calf, and the right lower extremity was unremarkable. Neurological examination was difficult to complete due to poor cooperation, but was grossly non-focal. An initial attempt at phlebotomy was unsuccessful. Abdominal radiography showed an ileus pattern with no free air under the diaphragm.

From: *Evidence-Based Medical Ethics*
By: J.E. Snyder and C.C. Gauthier © Humana Press, Totowa, NJ

The Ethical Dilemma

Betty's family presented her Living Will documentation which clearly stated a preference for a Do Not Resuscitate, Do Not Intubate (DNR/DNI) code status, and Betty's wish not to have a feeding tube placed for nutritional support in the setting of terminal, irreversible illness. When asked what their treatment expectations for Betty were, her family showed good understanding that Betty was at the end of her life. They wanted to know what could be done to make her comfortable in her "last days." They stated that in recent weeks, when Betty was more alert, she had stated that she wanted to "die peacefully" when "the time came." They requested assistance from the medical team in granting this wish.

> **Question for thought and discussion:** What can the medical team offer to Betty at this point in her illness?

> **Questions for thought and discussion:** Is giving Betty more pain medication ethical, knowing that it will likely cause further sedation and lethargy? What if the pain medication slows her respirations and hastens her death? Is this treatment the same as "assisted suicide?"

The Medicine

According to estimates from the National Cancer Institute (NCI) there will be 37,170 diagnoses of pancreatic cancer in the United States in 2007, and 33,370 will die from the disease. Incidence and mortality is slightly higher in men and in African-Americans. Overall, it is the fourth leading cause of cancer-related death in the United States, and the annual number of new diagnoses is increasing. The high mortality rate is due in part to the lack of screening methods that allow early disease detection. Around half of patients with pancreatic cancer are diagnosed when the disease has already metastasized, and there is an average life expectancy in these patients of less than six months. The five-year survival rate for patients diagnosed with any stage of pancreatic cancer is 5 percent. Risk factors for developing pancreatic cancer include advanced age, tobacco smoking, chronic pancreatitis, elevated body mass index, and late onset development of diabetes mellitus without usual risk factors. Additionally, a family history of pancreatitis, pancreatic cancer, or certain other heritable cancers increases risk of the disease.

Due to the limited number of treatment options for patients with any stage of pancreatic cancer, enrollment in clinical trials should be considered as one alternative to palliative care. Current recommendations for treatment of Stage IV pancreatic cancer (disease with distant metastases) is palliation, and includes the option of chemotherapy with gemcitabine or 5-fluorouracil. Response to standard chemotherapy is generally poor, with symptom relief in some patients, but limited survival benefit. Palliative stenting of blocked biliary ducts can be done endoscopically or percutaneously. Nerve blocks can also assist with pain management.

The Law

Betty may need to be admitted to the hospital so that her pain and other symptoms can be better controlled. Once she is admitted the medical team may consider palliative sedation for Betty that may require a change in pain medication or higher doses of the pain medication she is presently receiving. The medical team must understand that it is legal to provide enough medication to control pain, even if the medication also hastens death. This practice is also endorsed by the American Medical Association and the American College of Physicians.

The American Medical Association Code of Medical Ethics notes that, "Physicians have an obligation to relieve pain and suffering and to promote the dignity and autonomy of dying patients in their care. This includes providing palliative treatment even though it may foreseeably hasten death." Similarly, according to the American College of Physicians Ethics Manual, "with regard to pain control, the physician may appropriately increase medication to relieve pain, even if this action inadvertently shortens life."

The medical team must also understand that palliative sedation is not the same as physician-assisted suicide or euthanasia. In physician-assisted suicide the physician provides a prescription for lethal medication to a terminally ill patient who can then fill the prescription and take the medication to end his or her life. This is, as of the time of this book's publication, only legal in the state of Oregon. According to the Oregon "Death with Dignity Act," a state resident who is terminally ill (expected to die within six months), at least 18 years of age, and capable of making and communicating health care decisions, may "obtain a prescription for medication to end his or her life in a humane and dignified manner."

The patient must make an oral request, a written request, and a second oral request during a 15-day waiting period. The attending physician must confirm the patient's terminal illness and make sure the patient is capable and has made the request voluntarily. The patient must receive information about his or her diagnosis and prognosis, the potential risks and probable results of taking the medication, and alternatives such as hospice care and pain control. The patient must also see a second, consulting physician to confirm the diagnosis, ensure that the patient is capable and is making an informed and voluntary decision. If either of the physicians suspects that the patient is suffering from impaired judgment, a referral for counseling must be made.

Providing palliative sedation is physically different from giving a patient a prescription for medication to be used, by the patient, to end life. Many commentators also see a moral difference. For example, the late Supreme Court Justice William Rehnquist pointed out the difference in intention, arguing that, "when a doctor provides aggressive palliative care; in some cases, painkilling drugs may hasten a patient's death, but the physician's purpose and intent is, or may be, only to ease his patient's pain. A doctor who assists a suicide, however, 'must' necessarily and indubitably, intend primarily that the patient be made dead" (*Vacco v. Quill,* 1997).

Euthanasia occurs when a physician actually kills the patient (e.g., lethal injection). This practice, under guidelines similar to those in Oregon for physician-assisted suicide, is legal in the Netherlands. It is not legal anywhere in the United States. In both palliative sedation and euthanasia the physician physically injects the medication. However, many commentators point out the same moral difference as was noted by Justice Rehnquist in distinguishing palliative sedation and assisted suicide. In palliative sedation the purpose is to relieve suffering and hastening the patient's death is merely a side effect, while in euthanasia, the purpose is simply to end the patient's life.

The Ethics

In this case Betty's pain is not being fully relieved by the pain medication she is currently receiving. The family has brought Betty into the Emergency Department hoping that more can be done to make her comfortable in her last days. They explain that Betty wanted to "die peacefully" when "the time came" and they are requesting that the medical team help to make that possible. The medical team should begin by re-evaluating the medications Betty is currently taking for her pain and determine if they may need to give her something that is more effective in relieving her pain.

However, if whatever medication they prescribe is not working to provide Betty relief from pain, the medical team may consider palliative sedation. This would involve giving Betty enough medication to control her pain, although it may also hasten her death by depressing respiration. Even if the medical team understands that this would be legal and is permitted by the American Medical Association and the American College of Physicians, they may still have moral reservations about relieving Betty's suffering in a way that also has the side effect of hastening her death.

Applying the *Principle of Beneficence* to this case the medical team would consider the good promoted for Betty and the harm prevented by palliative sedation. Palliative sedation would prevent Betty's pain and suffering as she is dying from her terminal illness. It could also allow her to have a peaceful and humane death. With the *Principle of Non-Maleficence* the ethical question becomes more complicated. This principle requires medical practitioners to do no harm to their patients. Palliative sedation also has the side effect of hastening death and some may consider this to be a harm for her. On the other hand, because she is going to die soon no matter what medical treatment she receives – and she will be suffering while she dies – death may, in the end, be a good thing for Betty.

Combining these two principles, the medical team will need to determine what would be in Betty's best interests and compare the possibility of an earlier death with the relief of her pain as she dies. However, the consideration of these outcomes could be informed by Betty's own wishes.

According to the *Principle of Respect for Autonomy* medical practitioners should allow capable patients to make their own medical decisions, accepting or refusing recommended medical interventions. If the medical team is considering palliative sedation for Betty, they should also consider what she wanted in terms of how her life ended. Practitioners should not assume that a terminally ill patient is always incapable of making treatment decisions. Betty's family members have stated that when she was alert she expressed the desire to "die peacefully." This request should not be discounted because she is terminally ill, has probably been in and out of consciousness at various times as her cancer has progressed, and cannot make her own decisions at this point.

In many states, specific immediate family members are permitted to make medical decisions for the patient. Depending upon which family members are requesting a "peaceful death" for Betty, this could also be considered a legally-authorized request. Betty's spouse and adult children will be considered her legal surrogate decision makers. Moreover, they seem to be making the request for palliative sedation on the basis of both "substituted judgment," given Betty's own statement when she was alert, and her best interests, which would be to die without pain.

The *Principle of Respect for Dignity* is also relevant to this case. Betty has been living with family members during the end stages of her illness. This may indicate the value of her relationships with these family members and the trust she has placed in them as surrogate decision makers. Betty's own reasonable goals also need to be taken into consideration. When she was alert, Betty asked for a "peaceful death." She clearly does not want to endure any further invasions of her bodily integrity and she would prefer a death without unnecessary suffering. With palliative sedation the medical team could honor Betty's important relationships, meet her reasonable goals, and protect her from further bodily invasions and suffering.

The Formulation

Now that the evidence-based medicine, legal precedent, and relevant ethical principles for this case have been reviewed, formulate a strategy to address the ethical conflicts in this case. If necessary, perform additional research into local and state laws and hospital regulations. Consider delving further into the background medical literature to assist with making sound therapeutic decisions. Devise a treatment approach that addresses the needs of the patient and her family, that is both ethically and medically sound, and that is culturally competent. Ensure that the strategy employs fair and appropriate utilization of medical resources, and that the approach is practical and feasible within the limits of the medical system at large. Work out a clear and professional way to communicate the proposal to the patient and her family. Attempt to foresee challenges that may arise in conveying or implementing the plan. Determine what follow-up will be necessary to ensure that the chosen strategy remains successful for the patient in the long-term. Reflect on how the knowledge and skills learned from this case can be used to improve the care of patients that may be encountered in future practice.

Afterthoughts

In this case a patient presented to the hospital in the final stages of a terminal illness, in significant pain, and with little chance for the medical team to prolong her life. The dilemma was how to best manage the patient's comfort if doing so adequately meant a high likelihood of shortening the patient's life. In the realm of medical care, nearly all therapeutic options represent a choice where potential benefit must be balanced with potential risk. When does risk become too significant to even offer a treatment plan to a patient? Should a patient always be allowed to make the choice to receive a treatment, even in the setting of high risk? Can a provider ever refuse to administer the treatment in these situations? Does the provider's own personal belief system ever play into the decision-making process?

Annotated References/Further Information

Code of Medical Ethics of the American Medical Association: Current Opinions with Annotations, 2006–2007 Edition. Council on Ethical and Judicial Affairs. Annotations prepared by the Southern Illinois University Schools of Medicine and Law. The Subsection, "Withholding or Withdrawing Life-Sustaining Medical Treatment" (2.20) includes the AMA's opinion on palliative sedation.

Snyder L, JD, and Leffler C, JD. Ethics Manual, Fifth Edition. Ethics and Human Rights Committee, American College of Physicians. Annals of Internal Medicine. 142(7):560–582, 5 April 2005. In the subsection "Physician-Assisted Suicide and Euthanasia" the Ethics Manual makes the distinction between palliative sedation and both assisted suicide and euthanasia.

Ghaneh P et al. Biology and management of pancreatic cancer. Gut. 2007 Aug;56(8):1134–52.

The Pancreatic Cancer Action Network. Pancreatic Cancer Facts for 2007. Acquired from: *http://pancan.org/About/pancreaticCancerStats.html*. Accessed October 15, 2007.

National Cancer Institute. A Snapshot of Pancreatic Cancer. Acquired from: *http://www.cancer.gov/cancerinfo/types/pancreatic*. Accessed October 15, 2007.

National Cancer Institute. Stage IV Pancreatic Cancer. Acquired from: *http://www.cancer.gov/cancertopics/pdq/treatment/pancreatic/HealthProfessional/page9*. Accessed October 15, 2007.

Death with Dignity Act, 1997. Oregon Revised Statutes 127.800–995.

Vacco v. Quill (1997). U.S. Supreme Court. 521 U.S. 793.

Comprehensive Exam

1. **A 34-year-old male is evaluated in the Emergency Department for chest pain and is noted to have "track marks" on his arm, suggesting chronic injection drug use. The patient, however, denies the use of any illicit drugs. He signs a general consent form to receive medical care at the facility. Without obtaining more specific informed consent, the patient may now be tested for:**

   A. Drug use using a urine toxicology screen, to offer the patient substance abuse treatment options and to avoid possible interactions between medications and illicit drugs

   B. HIV infection, to offer the patient timely initiation of antiretroviral treatment if positive

   C. Both drug use and HIV infection

   D. Neither drug use or HIV infection

2. **A 32-year-old female is admitted to the hospital for gastroenteritis and volume depletion. After receiving intravenous fluids and medication to relieve her symptoms, she is able to eat a full meal without difficulty, and her abdominal cramping is only minimal. Her serum laboratory values are all within normal limits and her physician wants to discharge her. The patient requests to stay "one more night" in the hospital in case her symptoms return or pain becomes worse again. The best thing for her physician to do is:**

   A. To honor her request and discharge her the next day

   B. To honor her request, but tell her that the extra day of hospitalization will likely not be covered by her insurance

   C. To not honor her request and discharge her immediately with plenty of pain medication in case her symptoms return

   D. To not honor her request and discharge her immediately, but arrange an outpatient follow-up appointment for her and formulate a plan of action in case her symptoms recur at home

From: *Evidence-Based Medical Ethics*
By: J.E. Snyder and C.C. Gauthier © Humana Press, Totowa, NJ

3. **An unidentified man is brought to a Level 1 trauma center after being involved in a motor vehicle collision. He is unconscious and requires immediate surgery to stop intra-abdominal bleeding. Which of the following statements about informed consent for the procedure is the most appropriate?**

A. Informed consent is not necessary since it's an emergency procedure
B. Informed consent is not necessary since it's an emergency procedure, and brief documentation to this affect should be made in the medical record
C. Informed consent for the patient to have this procedure can only be given by a legal guardian, who is appointed by a judge, and legal proceedings for this should be done quickly given the urgency of the patient's medical condition
D. Informed consent is necessary for all procedures to ensure that human rights are protected

4. **A 65-year-old woman with widely metastatic breast cancer is admitted to the intensive care unit with intracerebral bleeding from her metastases. She is unsedated on a mechanical ventilator, and is not moving spontaneously or breathing above the set rate on the ventilator. When the medical team approaches her family to discuss her care, a family member states that "we know that God will help pull her through this." The best thing for the team to say to the family is:**

A. "Even God cannot help her now"
B. "Your expectations for her recovery do not sound realistic"
C. "Further medical care is futile at this point"
D. "May we involve the hospital chaplain in her care?"

5. **An eight-year-old girl undergoes an elective tonsillectomy at a hospital's outpatient surgery center. The procedure is done without any complications and afterwards, in the post-anesthesia care unit (PACU), the girl is awake and feeling well. Nonetheless, the girl's parents insist that she be admitted to the hospital for overnight observation. The surgeon feels that there is no medical justification for a hospital admission and that the girl requires no further skilled medical care. The best thing for the surgeon to do next is:**

A. Immediately discharge the girl home with her parents
B. Offer to send the girl home with a home nursing visit later in the evening
C. Observe the girl several more hours in the PACU, then discharge her home with her parents if she is doing well
D. Admit the girl to the hospital for overnight observation

6. **A 15-year-old girl comes to a routine gynecology visit and requests an HIV test from the physician. She reports that she has had voluntary, unprotected sexual intercourse with a 16-year-old male who has used intravenous heroin. She has not discussed this matter with her parents and does not want to. The best course of action for the physician to do is:**

   A. Test the patient for HIV infection now
   B. Test the patient for HIV infection now, counsel her about HIV risks and provide her with free condoms
   C. Test the patient for HIV infection now, counsel her about HIV risks, provide her with free condoms, and explore her reasons for not wanting to talk to her parents about the matter
   D. Test the patient for HIV infection now, counsel her about HIV risks, provide her with free condoms, explore her reasons for not wanting to talk to her parents about the matter, and offer to sit with her and her parents to talk through the matter

7. **An Emergency Department phlebotomist sustains a needlestick while caring for a patient. The patient refuses to be tested for HIV. Which of the following statements is the most correct?**

   A. In most states, the patient can be tested anyway since a caregiver was placed at risk
   B. The patient should not be tested for HIV antibodies, a CD_4^+-lymphocyte count, or an HIV viral load without informed consent
   C. The patient cannot be tested for HIV antibodies, but it is ethical to check a CD_4^+-lymphocyte count and HIV viral load without the patient's knowledge, since a caregiver was placed at risk
   D. The patient can still be tested for HIV infection, but the test results will not be reportable since the patient did not consent for the test

8. **A 76-year-old male with dementia is determined to not have the capacity to consent for an elective surgical procedure, and the patient's health care agent from their Power of Attorney for Health Care is unable to participate in the decision for personal reasons. From which of the following persons is it most appropriate to obtain consent?**

   A. The patient's neighbor, who has known him well for 40 years
   B. The patient's first cousin, who hasn't spoken to the patient in two years
   C. The patient's alternate health care agent choice, as listed on the Power of Attorney for Health Care, who hasn't spoken to the patient in four years
   D. No one is able to give consent for the patient to have the procedure in this situation

9. **A 25-year-old patient is admitted to the hospital with fever and cellulitis from injection drug use. The medical team, also concerned about an auscultated heart murmur, wants to order an echocardiogram to rule out endocarditis. While in his hospital room, the patient is found by his nurse to be using injection heroin with a visitor. Which of the following is the best choice for his medical team to do next?**

A. Discharge the patient immediately to prevent an adverse event on hospital premises such as a heroin overdose

B. Discharge the patient immediately to prevent an adverse event on hospital premises such as a heroin overdose, but provide him with a prescription for antibiotics, orders for an outpatient echocardiogram, and a follow-up appointment

C. Continue to treat the patient, using hospital police to confine him to his hospital room and not allow him any further visitors

D. Continue to treat the patient, but arrange a verbal or written contract with him that clarifies agreed-upon treatment goals and rules for behavior while he is in the hospital

10. **A 31-year-old patient with a history of schizophrenia presents to the Emergency Department with a subcutaneous abscess of the left upper extremity. When the surgical resident attempts to obtain consent to perform an incision and drainage, the patient refuses. How can it be determined if the patient has the capacity to consent for or refuse to have this procedure?**

A. If the patient is alert and oriented to person, place, and time, then he has the capacity to consent to or refuse the procedure

B. If the patient demonstrates an understanding of the risks and benefits of both having the procedure and refusing the procedure, then he has the capacity to make the decision about having the procedure

C. The capacity to give consent by this patient may only be determined by a psychiatrist

D. Consent is not necessary in this situation because the incision and drainage is an emergency procedure

11. **An 87-year-old male with dementia is admitted to the hospital with chest pain, and becomes acutely confused and agitated during the first night of his hospitalization. He is yelling loudly and hits a nurse who was trying to calm him down. When the nurse tried to apply soft wrist restraints to safely provide him with care, he yells "no, I don't want those!" The best thing for his medical team to do is:**

A. Continue to try and use verbal redirection to calm him down

B. Call a psychiatry consult

C. Heavily sedate him with medications, and use physical restraints to prevent him from harming hospital staff

D. Use chemical and physical restraints temporarily to calm him down so that treatment can be safely initiated, and re-evaluate his medical status and the need for continued restraints on a regular basis

12. **A 31-year-old man is involved in a motor vehicle collision and sustains severe bleeding from a laceration. He arrives at the hospital tired and pale, but conversive and appropriate. Laboratory evaluation shows a marked anemia. Further history-taking determines that he is a Jehovah's Witness. The physician caring for this patient should *not*:**

   A. Have the patient's family briefly step out of the room and ask the patient in private what his thoughts regarding transfusion are

   B. Provide the patient with information about the possible morbidity and mortality associated with refusing blood transfusion in this situation

   C. Advise the patient to consider contacting his religious leaders for help in deciding whether to receive a blood transfusion or other products

   D. Suggest a Do Not Resuscitate/Do Not Intubate (DNR/DNI) code status based on his religious convictions

13. **A 51-year-old male with a remote history of alcohol and intravenous drug abuse, and currently with chronic hepatitis C and mild hypertension, requests information from his primary care provider about antiviral therapy. He has been sober for over 10 years, and is adherent to his one anti-hypertensive agent and all scheduled medical visits. Treatment for hepatitis C will require routine medical visits and blood work. Which of the following statements is the best reply for his primary care provider to make?**

   A. "Your past history of drug and alcohol abuse does not preclude you from potentially receiving antiviral therapy"

   B. "Your past history of drug and alcohol abuse does not preclude you from potentially receiving antiviral therapy, but you will need to demonstrate continued adherence to treatment, lab work, and medical visits in the future"

   C. "Your past history of drug and alcohol abuse does not preclude you from potentially receiving antiviral therapy, but you will first need to undergo a psychiatric evaluation"

   D. "Your past history of drug and alcohol abuse precludes you from receiving antiviral therapy"

14. **A 59-year-old man is diagnosed with advanced stage pancreatic cancer and requests a lethal dose of morphine from his primary care physician. The best response for the physician to give is:**

   A. "There are many ongoing trials for chemotherapy agents that can extend your life"

   B. "Am I not doing a good job controlling your pain and other symptoms?"

   C. "I will write you the prescription, but you can't tell anyone where you got the medication from"

   D. "Physician-assisted suicide is only legal in one state; you will have to seek that treatment there"

15. A 51-year–old-man is newly diagnosed with lung cancer. He comes to a medical appointment with his male partner to discuss treatment options. In addition to providing information on the treatment of the lung cancer, the provider should:

A. Determine what his wishes would be with regard to life-sustaining treatments, and document these in the medical record

B. Advise the patient to complete a Living Will document where he expressly states his wishes with regard to life-sustaining treatments

C. Advise the patient to complete a Power of Attorney for Health Care document to name a surrogate decision maker for himself, in case he becomes unable to make his own health care decisions at some point

D. Do all of the above

16. A 24-year-old man without significant past medical history appears for a routine health visit with a new primary care physician. At the appointment he makes several sexually inappropriate remarks to the physician, as well as several other vaguely threatening comments. The physician feels too uncomfortable to continue caring for the patient in the future. The best thing for the physician to do is:

A. Discharge the patient from the medical practice immediately

B. Give the patient a written letter that discharges him from the medical practice, but that also offers choices for emergency treatment until another provider can take over the patient's care

C. Offer to continue seeing the patient in the future, so as to not commit patient abandonment

D. Offer to continue seeing the patient in the future, so as to not commit patient abandonment, but do so with a security guard present at all times

17. A patient demands to have a CT scan of her spine to evaluate her lower back pain. In the clinical judgment of her physician, the pain is due to a muscle spasm; therefore, he believes a CT scan is unnecessary. The best thing for the physician to do is:

A. Order the CT scan so that the patient will not file a malpractice law suit

B. Order the CT scan, but tell the patient that insurance is not going to pay for it since it is not medically indicated

C. Do not order the CT scan, but offer to refer the patient to another physician for a second opinion

D. Do not order the CT scan and tell the patient that she will need to seek medical care elsewhere

18. A 98-year-old man is admitted to the hospital with chest pain. The admitting physician wants to initiate a discussion with him about code status. Which of the following statements is the most appropriate?

A. The patient, if capable of deciding, should be asked what his wishes are regarding resuscitation after risks and benefits are explained
B. Because of his advanced age, code status should be determined based on the patient's pre-hospital quality of life
C. Resuscitation would not be an appropriate use of medical resources in this case, as 98-year-old patients generally do not survive resuscitation
D. The physician should not be discussing code status with a 98-year-old patient

19. **A 38-year-old man with end-stage chronic kidney disease and anemia of chronic disease agrees to begin hemodialysis therapy. The nephrologist wishes to begin erythropoietin therapy for his anemia, but the patient refuses this treatment and wants to take iron supplements instead. The physician explains that iron is not an appropriate treatment for his condition, but the patient refuses to change his mind. The best thing for the physician to do is:**

A. Prescribe only the medications that the patient agrees to take
B. Refuse to offer him any further treatment, since his choices are illogical and are not compatible with the medical standard of care
C. Delve further into the reasons for his choices and make a determination about his capacity to make decisions about treatment
D. Refer him to a psychiatrist

20. **A 70-year-old patient has no legal documentation such as a Living Will or Power of Attorney for Health Care, and loses the capacity to make decisions about her own medical care. She is widowed and has five adult children, three of whom want her to have an elective procedure and two that don't. The decision about having the procedure should:**

A. Be determined by the children with the majority opinion
B. Be determined only when unanimity has been reached among the children
C. Be determined by bringing in the patient's elderly mother and counting her vote
D. Be determined by an objective third party, such as a legally appointed guardian

21. **During a routine health visit with a new primary care physician, a 37-year-old woman with chronic lower back pain requests narcotic analgesics for her symptoms. The physician states that he will provide a limited number of these medications to the patient on a month-by-month basis if she agrees to sign a narcotic contract that has several stipulations, including routine urine toxicology screens and that she cannot receive narcotics from other prescribers. The patient refuses to sign the contract and demands that the physician provide her with the medications anyway. At this point, the physician should:**

A. Offer to refer the patient to a physician who is a pain management specialist to further evaluate her symptoms
B. Agree to prescribe the patient a limited amount of narcotics just this one time

C. Agree to prescribe the patient a limited amount of narcotics monthly if she verbally agrees to use them only as directed

D. Discharge the patient from his medical practice

22. **A 44-year-old male is admitted to the hospital with acute pancreatitis due to alcohol abuse, and is made *nil per os* (NPO), and given intravenous fluids and analgesics. He reports to the medical team that his pain is not getting better over the first several days of his hospitalization. A CT scan of the abdomen is ordered to evaluate him for complications of his pancreatitis, but is delayed several times since the patient is out of his hospital room every time the radiology transporter comes to get him for the test. His nurse suspects that he is leaving his room to eat at the hospital coffee shop. The best thing for his medical team to do is**

    A. Arrange a verbal or written contract with the patient that clarifies agreed-upon treatment goals and rules for behavior, such as being available in his hospital room at certain times when care must be given and adherence to his NPO order

    B. Discharge the patient home since he is not following orders from the medical team

    C. Tell the patient that he must remain in his hospital room at all times, or he will be discharged home

    D. Have a nursing assistant keep the patient under constant observation to assure adherence with his orders

23. **Patient A is critically ill and requires numerous treatments with Medication X to be kept alive. Medication X is very expensive, and only a limited number of doses are available at the hospital. Two other patients, Patient B and Patient C, are then admitted to the hospital and will also need Medication X in order to be kept alive. Which of the following is the best justification to stop giving Medication X to Patient A?**

    A. The total cost assumed by giving Medication X to Patient A

    B. The relatively low contribution to society that Patient A offers, compared to the higher contributions of Patients B and C

    C. The likelihood that all three patients will die if Patient A continues to receive Medication X and there is not enough of it to give to Patients B and C

    D. Patient A is much older than Patient B and Patient C

24. **While cleaning up his desk, a primary care physician finds test results that are two months old indicating one of his patients has hyperthyroidism. The patient's next appointment with the physician is in three weeks. The next best step for the physician to take is to:**

    A. Apologetically disclose all information about how the error occurred in person at the patient's next scheduled appointment

B. Immediately arrange an earlier office appointment with the patient to personally notify her about the test results, and to apologetically disclose all information about how the error occurred

C. Immediately send the patient an official letter advising her about the test results that apologetically discloses all information to her regarding the error, and then keep the next scheduled appointment with her in three weeks

D. Immediately have his attorney send the patient a certified letter advising her about the test results that apologetically discloses all information to her regarding the error, and then keep the next scheduled appointment with her in three weeks

25. **A 42-year-old woman presents to the hospital with a large subdural hematoma that requires emergent surgical evacuation. The medical team caring for her suspects that the hematoma is a result of domestic violence. The patient is not capable of consenting for the procedure due to an altered level of consciousness. The best person to give consent for this procedure is:**

A. The patient's husband, despite the possible abuse, since they are legally married

B. The patient's mother, although she lives two hours away, since there is a possibility that the patient's husband caused the injury

C. A surrogate decision maker determined by the state, since there is a possibility that the patient's husband caused the injury and he legally has more rights than the patient's mother or other relatives

D. No one, since the procedure is an emergency one and informed consent is not required

26. **A 31-year-old man dies from complications of an AIDS-related pneumonia. His wife requests that HIV and AIDS not be listed as the primary cause of death on the death certificate, since this is a document of public record and she doesn't want other people to know about his illness. The best thing to do in this situation is to:**

A. List "AIDS" and "AIDS-related pneumonia" as the primary cause of death on the death certificate

B. List "Unknown," but do *not* list "AIDS" or "AIDS-related pneumonia" as the primary cause of death on the death certificate since the patient's wife is legally the patient's health care agent and her request must be honored

C. List "Unspecified pneumonia," but do *not* list "AIDS" or "AIDS-related pneumonia" as the primary cause of death on the death certificate since the patient's wife is legally the patient's health care agent and her request must be honored

D. Refer the case to the medical examiner to better determine the primary cause of death

27. **A 27-year-old man is admitted to the hospital with chest pain that started one hour after using crack cocaine. Urine toxicology confirms his cocaine use. The man reports that he occasionally sells drugs to support his own cocaine use. He is the father of an eight-year-old girl who lives with him and his girlfriend every other weekend, and with his ex-wife at all other times. The medical team should report his drug use to:**

   A. No one since this information must be kept confidential
   B. The local department of social services so that his daughter's welfare and safety can be investigated
   C. Local law enforcement since his activity is illegal and he should be prosecuted
   D. His ex-wife so she can seek sole custody of their daughter and keep her safe

28. **An unidentified and unconscious woman, likely in her late 30s, is brought to the Emergency Department and found by laboratory evaluation to be in a diabetic hyperosmolar non-ketotic state. Attempts at peripheral intravenous access are unsuccessful, due to volume depletion and hypotension. Which of the following statements about informed consent for placing a central venous access is the most appropriate?**

   A. Informed consent is necessary for all procedures to ensure that human rights are protected
   B. Informed consent is not necessary as the procedure in this case is an emergency one
   C. Informed consent is not necessary as the procedure in this case is an emergency one, and brief documentation to this affect should be made in the medical record
   D. Informed consent for the patient to have this procedure can only be given by a legal guardian, who is appointed by a judge, and legal proceedings for this should be done quickly given the urgency of the patient's medical condition

29. **Which of the following laboratory tests is usually not included under the blanket category of general consent for medical care when a patient is admitted to the hospital?**

   A. CD_4^+-lymphocyte count
   B. HIV viral load
   C. Hepatitis C antibodies
   D. Certain genetic assays

30. **A 79-year-old African-American male has metastatic lung cancer, with bone and brain involvement. He presents to the hospital with mental status changes, and the admitting physician asks the family what his wishes would be with regard to treatment in the event of a cardiopulmonary arrest. They state that they want "everything to be done." The best thing for the physician to reply is:**

A. "I don't think that resuscitating him is an appropriate use of medical resources, or truly in his best interest"

B. "He has metastatic cancer. I don't think that resuscitation is going to make a difference in helping him live longer"

C. "Based on good medical evidence, the likelihood of him surviving a cardiopulmonary arrest is very low with or without resuscitation, and I'm concerned that resuscitating him may actually do him more harm than good"

D. "I think that you may relate better to an African-American physician, and I'd be happy to refer you to one if you'd like"

31. A 72-year-old patient with a terminal condition is not capable of giving consent for a life-sustaining procedure. Which of the following is the appropriate person to obtain consent from?

A. The patient, per their specific wishes about such a procedure as stated in their Living Will

B. The person who is named as the health care agent on the patient's Power of Attorney for Health Care paperwork

C. The patient's legal spouse

D. A surrogate decision maker who is determined to know the patient well

32. A 50-year-old man with HIV infection suffers a devastating stroke and is in the intensive care unit on a mechanical ventilator. He had not previously completed any legal documentation such as a Living Will or a Power of Attorney for Health Care. The closest living relative to make health care decisions for him is his father, who has not spoken to the patient in two years, and the medical team does not know if he is aware of his son's HIV diagnosis. The medical team caring for the patient should:

A. Be completely honest about the patient's medical condition with his father, including disclosure of the HIV diagnosis, so that he can make informed decisions about his son's care

B. Be completely honest about the patient's medical condition with his father so that he can make informed decisions about his son's care, but not disclose the HIV diagnosis since it is not relevant to care of the patient's stroke

C. Determine how much the father knows about his son's past medical problems before any further discussions; then be completely honest about the patient's medical condition so the father can make informed decisions about his son's care, but the HIV diagnosis should not be newly disclosed if not previously known since it is not relevant to care of the patient's stroke

D. First determine if the father is an appropriate surrogate decision maker for the patient, given that the father and son haven't spoken in two years

33. **A 100-year-old woman is admitted to the hospital with rectal bleeding. With regard to performing a colonoscopy, which of the following statements is the most correct?**

    A. Patients that are 100-years-old should not have invasive procedures done
    B. A colonoscopy should only be done if the patient is a vibrant and active 100-year-old
    C. A colonoscopy should only be done if the patient is able to have the procedure covered by Medicare or private medical insurance, or be paid out of her personal savings
    D. The patient, if capable of deciding, should be asked if she wants a colonoscopy, after risks and benefits are explained

34. **A 65-year-old man is admitted to the hospital with acute myocardial infarction. The admitting physician initiates a conversation about code status with the patient and his family. The patient's wife asks the physician if she can be present during resuscitation if the patient has a cardiopulmonary arrest. The best reply to her would be:**

    A. "No, I think that it would be best if you were not present"
    B. "Yes, the family has the right to be present for the resuscitation"
    C. "Yes, a limited number of adult family members may be present for the resuscitation, but they must allow providers to care for your husband without disruption"
    D. "Yes, a limited number of adult family members may be present for the resuscitation, but they must be prepared to see some potentially gruesome things"

35. **If a patient is refusing to have a recommended procedure performed, and the procedure is one that the treating physician feels is both standard of care and potentially lifesaving, the physician must:**

    A. Assume that the patient does not have the capacity to decide whether to have the procedure or not
    B. Keep trying to convince the patient that the procedure is necessary until they eventually consent for the procedure
    C. Obtain a psychiatric consult on the patient
    D. Do not perform the procedure if the patient is determined to have the capacity to make this decision

36. **A 47-year-old woman has poorly controlled Type 2 diabetes mellitus and her physician wants to begin outpatient treatment with insulin. Her religious views require that she fast during the daytime for one month each year. Which of the following statements is the most appropriate with regard to her diabetes treatment?**

    A. Insulin is not a good choice of medication for this patient due to her dietary variability and high risk of developing hypoglycemia during fasting

B. The physician should advise the patient to seek an exemption for the fasting from her religious leader

C. The patient should not receive insulin treatment during her fast unless she, unprompted, offers to seek an exemption for the fasting from her religious leader

D. The patient should be referred to another physician who shares her religious views

37. **A baby girl requires extracorporeal membrane oxygenation (ECMO) immediately after birth, due to birth defects and respiratory dysfunction. On day 27 of treatment on the ECMO machine, at a cost of several thousand dollars per day, the girl remains critically ill and has few positive signs of recovery. Which of the following statements is most correct?**

A. Further health care dollars should probably not be spent on this girl's care

B. Due to the lack of improvement in the girl's condition, and the low likelihood that she will live to become a contributing member of society, there should be consideration of stopping the ECMO

C. Since there is only one ECMO machine at this hospital, the girl should be removed from the machine if another critically ill infant suddenly needs it

D. If another critically ill infant suddenly needs an ECMO machine, then that patient should be transferred to a nearby medical facility that has a machine available

38. **A 27-year-old man is newly diagnosed with HIV. He admits to his primary care physician that his wife is unaware of his diagnosis and that he is still having unprotected sexual intercourse with her. He states he is too ashamed to tell her about his diagnosis and how he acquired the infection, and asks the physician not to tell her either. Of the following, the physician's best reply is:**

A. "I will keep your HIV diagnosis just between us since everything you tell me is confidential"

B. "I will keep your HIV diagnosis just between us since everything you tell me is confidential, but you should really consider telling her the truth"

C. "I am personally required by law to tell the health department of your diagnosis, and they may notify your wife that she is at risk for contracting HIV from someone she is in a relationship with"

D. "I am personally required by law to have you arrested now for knowingly putting your wife at risk for contracting HIV infection"

39. **A 34-year-old male with a past medical history of bipolar disorder with psychotic features, alcohol abuse, and medication non-adherence presents to the Emergency Department with acute psychosis and has suicidal and**

homicidal ideations. He is screaming loudly and threw a punch at one of the nurses. The best thing for his medical team to do next is:

A. Heavily sedate him with medications and use physical restraints to prevent him from harming hospital staff

B. Take him to a room where he will not disturb other patients and, if necessary, use chemical and physical restraints temporarily to calm him down so that treatment can be safely initiated

C. Continue to try and use verbal redirection to calm him down

D. Call a psychiatry consult

40. **Which of the following statements about HIV-related testing is *not* correct?**

A. Positive HIV test results are always reported to the Centers for Disease Control and Prevention (CDC)

B. In a few states HIV testing can be done confidentially so that positive results are not reported to the Centers for Disease Control and Prevention (CDC)

C. In most states low CD_4^+-lymphocyte count test results are reported to the Centers for Disease Control and Prevention (CDC)

D. In most states positive HIV viral load test results are reported to the Centers for Disease Control and Prevention (CDC)

41. **A 51-year-old woman is started on cholesterol-lowering medication by her primary care physician based on the results of a fasting lipid profile. The physician later is notified by the testing laboratory that blood samples were accidentally mixed up, and that the patient's correct test results indicate that she did not have high cholesterol at all. The best action for the physician to take is to:**

A. Immediately advise the patient by phone to stop taking the medication, and to apologetically disclose all information to her regarding the error

B. Send the patient an official letter advising her to stop taking the medication, and to apologetically disclose all information to her regarding the error in the letter

C. Have his attorney send the patient a certified letter advising her to stop taking the medication, and to disclose all information to her regarding the error in the letter

D. Require the laboratory to immediately notify the patient about their reporting error, and then send her a letter of apology afterwards with directions to stop taking the medication

42. **A 27-year-old woman with HIV infection is offered antiretroviral therapy by her infectious disease specialist. The woman agrees to take nucleoside analog reverse transcriptase inhibitors (NRTIs) and non-nucleoside**

reverse transcriptase inhibitors (NNRTIs), but refuses to take protease inhibitors (PIs). Her doctor should:

A. Prescribe only the medications that she agrees to take
B. Refuse to offer her further treatment, since her choices are illogical and are not compatible with the standard of care for HIV
C. Ask about the reasons for her choices and determine if she is capable of making decisions about antiretroviral therapy.
D. Refer her to a psychiatrist

43. A 75-year-old woman with widely metastatic breast cancer is admitted to the hospital for pain control at the end of her life. She is lethargic, has shallow respirations, but is grimacing and moaning. The patient's son requests that she receive a higher dose of morphine than she is currently getting. The physician caring for this patient should:

A. Explain that you cannot give her more morphine since the higher dose may hasten her death
B. Explain that you cannot give her more morphine since the higher dose may hasten her death, and report the son's request to law enforcement officers
C. Give her more morphine, but explain to her son that the higher dose may hasten her death
D. Give her enough morphine so that she passes away more quickly

44. A 47-year-old woman with poorly controlled diabetes, who previously lived alone in a second floor apartment, is admitted to the hospital with a non-healing foot infection. She is diagnosed with osteomyelitis and undergoes a left below-the-knee amputation. Her pain is well controlled postoperatively and when she is medically ready for discharge she refuses several choices of skilled nursing facilities that would care for her incision site and provide rehabilitation therapy to help her become mobile again. The best discharge plan for her at this time is:

A. To offer to have an ambulance bring her to her apartment
B. To keep her in the hospital until she agrees to go to a skilled nursing facility
C. To determine why she is not willing to go to the facilities offered, and what she would like to do instead
D. To call her relatives, explain her medical situation and have them convince her to go to a skilled nursing facility

45. If an unmarried adult patient has no legal documentation such as a Living Will or Power of Attorney for Health Care, and loses the capacity to make decisions about their own medical care, in most states which of the following persons legally would become the surrogate decision maker?

A. The patient's adult child

B. The patient's adult sibling

C. The patient's parent

D. The patient's best friend, if they are an adult

46. A 52-year-old man is newly diagnosed with hepatocellular carcinoma, and prognosis is deemed to be poor. His primary physician should:

A. Determine what his wishes would be with regard to life-sustaining treatments, and document these in the medical record

B. Advise the patient to complete a Living Will document where he expressly states his wishes with regard to life-sustaining treatments

C. Advise the patient to complete a Power of Attorney for Health Care document to name a surrogate decision maker for himself, in case he becomes unable to make his own health care decisions at some point

D. Do all of the above

47. A 51-year-old woman with peripheral arterio-occlusive disease and a history of aorto-bifemoral bypass grafting two years previously presents to her vascular surgeon with symptoms of gradually worsening lower extremity claudication. Since her prior surgery she has continued to smoke cigarettes and is frequently non-adherent with her prescribed medications. Her surgeon believes that her non-adherence with prior recommendations has resulted in a rapid recurrence of arterio-occlusive disease. Repeat surgery is considered a procedure of moderate risk to the patient, given her general underlying condition. Before the vascular surgeon should consider performing a repeat surgery to alleviate the patient's symptoms, the patient should:

A. Have her capacity to decide on the surgery evaluated by a psychiatrist

B. Demonstrate a reasonable attempt to quit smoking and adhere to medical therapy

C. Seek a second opinion about the etiology of her symptoms

D. This patient should not have repeat surgery offered to her due to a history of non-adherence to medical recommendations

48. A 16-year-old girl presents to her pediatrician for a routine health visit. When her mother steps out of the room, the girl admits to being sexually active and having missed her last menstrual period. She requests a pregnancy test and the results are positive for pregnancy. She states that she does not want her mother to know about her pregnancy until she has had time to think about what she wants to do. When the mother returns to the examination room she asks what was discussed in her absence. Of the following, the best reply from the physician is:

A. "Your daughter's medical information is confidential"

B. "You should ask your daughter what we discussed"

C. "We discussed various aspects of your daughter's health, and everything is fine"

D. "I think we should all sit down together and talk about what we found out"

49. A 77-year-old woman is evaluated for a lump in her breast that is later determined by biopsy to be malignant. Before the patient is notified of the results, the patient's daughter calls the physician to find out the diagnosis. She states "if it's cancer, please don't tell her the diagnosis until after her granddaughter's wedding next month; it will just devastate her." In this case, the physician should:

A. Arrange the first available medical visit for the patient after the wedding to relate the cancer diagnosis to the patient

B. Arrange for the patient to immediately meet with an oncologist and grief counselor so they can relate the cancer diagnosis to the patient

C. Explain to the daughter that the patient does have cancer and that he is obligated to share the diagnosis with the patient herself as soon as possible

D. Explain to the daughter that he is obligated to share the biopsy results with the patient herself as soon as possible, and not specifically disclose the diagnosis to the patient's daughter

50. A 41-year-old married woman is diagnosed with gonorrhea. She tells her provider that she has had extramarital affairs and that she does not want her husband to know about the infection. She states that she has not been sexually active with her husband for over a year. The best thing for the provider to tell her is:

A. "I am required by law to tell your husband about your gonorrhea"

B. "I am not required by law to tell your husband about your gonorrhea, but you should strongly consider telling him yourself"

C. "I am not required by law to tell your husband about your gonorrhea, but someone from the Department of Health may notify him"

D. "I think that you should consider a divorce"

Comprehensive Exam — Answer Key

1. **A.** In most states HIV testing requires specific informed consent, but drug screens are considered covered by the general consent for medical care. The general practice of ordering urine drug screens, for medical purposes only, is considered legally acceptable. However, any use of drug screens for purposes other than providing safe medical care, particularly if done without the patient's knowledge, is not appropriate. Please refer to Case 7 for further discussion about general consent for medical care and informed consent for drug testing.

2. **D.** Health care resources must be used appropriately, and this patient no longer meets the criteria for inpatient care. Outpatient follow-up seems to be a reasonable compromise in this case. The prescription of extra pain medication is improper. Please refer to Case 3 for further discussion about criteria for hospitalization, and also to Case 8 for further discussion about the rights of providers to discharge patients who don't meet criteria for further hospitalization.

3. **B.** Informed consent is preferable, but not always necessary in the case of life-threatening emergencies. Documentation in the medical record is essential when a procedure is performed on a patient without their consent. Waiting for a legal judgment is not appropriate in an emergency situation. Please refer to Case 15 for further discussion about consent in patients who are unidentified and lack capacity to make decisions about medical care.

4. **D.** It is essential to respect the points of view of patients and their families from different cultures and religious backgrounds, as they may have different belief systems than the provider. Bridging cultural gaps will most likely help give patients the best possible care. Providing appropriate support and using an evidence-based approach may assist the provider in this process. Please refer to Case 13 for further discussion about futile care and the influence of culture on decision making at the end of life.

5. **A.** Health care resources must be used appropriately, and this patient no longer meets the criteria for inpatient care. Likewise, the use of busy PACU or nursing staff is not appropriate, and takes their time away from patients with more

critical needs. Please refer to Case 3 for further discussion about criteria for hospitalization, and also to Case 8 for further discussion about the rights of providers to discharge patients who don't meet criteria for further hospitalization.

6. **D.** Patients who are legal minors can often make their own decisions about health care without involving their parents. The Code of Ethics of the American Medical Association suggests that providers delve into why minors do not want their parents involved in their care, and to encourage parental involvement when appropriate. This patient additionally needs counseling on reducing her risk behaviors with regard to HIV infection and other sexually transmitted diseases. Please refer to Case 24 for further discussion about the rights of minors to consent for medical care.

7. **B.** Although the law in most states dictates that HIV testing requires informed consent, and no laws exist regarding formal consent for testing of surrogate laboratory markers for HIV, results from these surrogate markers may still be reportable. The most ethical thing to do is notify the patient, and preferably obtain their consent, about testing for HIV-related markers. Positive test results are reported, regardless of the status of patient consent. Please refer to Case 19 for further discussion about informed consent for HIV testing.

8. **C.** When a patient does not have the capacity to make a health care decision on their own, the health care agent from the Power of Attorney for Health Care is the highest on the hierarchy for surrogate decision making of the listed choices. This individual is the person who the patient felt, at a time when he had capacity, could best make decisions for him in just such a situation. Clearly, since the patient hasn't spoken to the alternate health care agent in some time, the ability of the health care agent to make appropriate medical decisions for the patient must be analyzed by the practitioner. Please refer to Case 2 for further discussion about when a primary choice for a health care agent cannot fulfill their obligations, and also Chapter 3 for a general discussion about Advance Directives.

9. **D.** Ensuring that safe and efficient care is provided to this patient is a priority and, unfortunately, some behavioral "ground rules" need to be set. Discharging the patient is not a good choice, since he has a potentially deadly diagnosis, and outpatient treatment is not standard of care. Confinement by the police is too restrictive of the patient's rights. Please refer to Case 5 for further discussion about patients whose behavior limits the ability of their providers to care for them, and the ethics of behavioral contracts.

10. **B.** Patients with mental illness may have the capacity to make health care decisions for themselves. They must demonstrate capacity to make such decisions in the same way as patients without mental illness, which relies on more information than level of alertness or orientation. The risks and benefits of accepting

or refusing the procedure must be explained to and understood by the patient. A psychiatry consult may be helpful in this case, but is not necessary in determining the patient's capacity to make this decision. At this time, the abscess does not appear to be life-threatening, so informed consent should be obtained from someone. Please refer to Case 1 for further discussion about determining a patient's capacity to give consent to procedures.

11. **D.** The use of chemical or physical restraints is sometimes necessary in the care of patients to reduce harm to the patient and/or the health care providers. The minimal amount of restraint should be used and the need for continued restraint re-evaluated frequently. Waiting for a psychiatry evaluation does not seem practical at this time due to the acute level of patient agitation and his behaviors, and verbal redirection has already failed to calm the patient. The patient's refusal of the restraints is done when he does not have the capacity to decline this intervention or his general care. Involving a loved one may help to calm the patient down, and that individual may be able to assist in making decisions for his care. Please refer to Case 6 for a further discussion about the use of physical and chemical restraints on patients.

12. **D.** The physician should not assume that there are no options available to treat this patient. The patient should be asked, in private and without coercion or duress, what treatments he will or will not accept. Spiritual guidance and the use of evidence-based medicine may help him make decisions in this regard. Please refer to Case 23 for a further discussion about caring for Jehovah's Witnesses.

13. **B.** Receiving antiviral therapy is likely in this patient's best interest, and his years of sobriety and adherence to medical recommendations potentially make him a good candidate for this treatment. Continued adherence will be important to the success of the treatment and in reducing risks associated with the treatment. There is no indication for a psychiatry evaluation at this time. Please refer to Case 20 for further discussion about the requirements of patients to prove adherence to medical therapy before certain treatments are made available to them.

14. **B.** Physician-assisted suicide is currently illegal in the majority of states. The reasons for this patient wanting to die quickly may be related to poor control of pain, depression, or other conditions, and the physician should inquire about these. The patient does not seem to be interested in extending the length of his life at this time, although trials of palliative therapy are worth exploring after the reasons for his primary request are thoroughly addressed. Please refer to Case 25 for further discussion about palliative sedation and physician-assisted suicide.

15. **D.** Patients with potentially terminal conditions should have their wishes regarding end-of-life care clearly documented in the medical record, and

supplemented with legal documentation when possible. This patient should determine who he would want to make health care decisions for him if he loses the capacity to do so. If he wishes for his partner to be his health care agent, then he should consider legal documentation to designate him as such since many states do not give same-sex partners legal rights in these situations. His wishes regarding resuscitation, mechanical ventilation, and nutrition are some of the other parameters that should be clarified. These discussions are often difficult, particularly in patients with newly diagnosed terminal illness, and sensitivity must be used while initiating them. Please refer to Cases 11 and 16 for further discussions about surrogate decision makers, Case 11 on the rights of same-sex partners in medical decision making, and Chapter 3 for a general discussion of Advance Directives.

16. **B.** Physicians should not put themselves at risk while caring for patients, particularly if they feel unsafe around them. Having a security guard present likely violates a patient's privacy, particularly as sensitive patient information is protected by the federal Health Insurance Portability and Accountability Act (HIPAA) Privacy Rule. A patient may be discharged from a medical practice, but they must be provided with alternative means of emergency care until another practitioner can assume their care. Please refer to Case 9 for further discussion about administratively discharging a patient from a medical service.

17. **C.** Practicing "defensive medicine" by ordering unnecessary tests is both inappropriate utilization of health care resources and needlessly expensive. The CT scan should not be ordered for this patient if it is not indicated, regardless of their request. If the practitioner is providing at least the standard of care to this patient, then a malpractice claim should not be of concern. There is not a clear reason to discharge the patient from the medical practice completely, and to do so may be construed as patient abandonment. The best choice in this case is to offer the patient an opportunity to get a second opinion on her symptoms. Please refer to Case 3 for further discussion about patients coercing providers into unnecessary tests and procedures, and also to Case 9 for further discussion about administratively discharging a patient from a medical service. Chapter 1 additionally has a discussion about the practice of "defensive medicine."

18. **A.** It is appropriate to discuss code status with a 98-year-old patient, and truthfully most all adult patients, regardless of age or current health status, should think about what their wishes would be at the end of life or in emergency situations. Although it is true that this patient is beyond the current life expectancy of a person in the United States, decisions about resuscitation should be made on an individual basis and not based on a physician's preconceived notions of what an older patient might want or what their pre-hospital quality of life might be. This is true even if providing resuscitation for the patient may not significantly prolong the duration of their life. It is an entirely different matter if

resuscitative efforts are known for a fact to be futile due to the patient's underlying illness – in this situation, resuscitative efforts probably should not be offered at all. Withholding medical resources from a patient generally should not be done on an individual basis, particularly if those resources are plentiful. Please refer to Case 14 for further discussion about the influence of age on providing certain medical care. Please also see Case 22 for a review of health care resource utilization and limitations thereof.

19. **C.** A provider should not offer alternative treatments that are not standard of care just because a patient requests them. In this case there is not a clear reason to withhold all treatments from the patient, since the patient is agreeable to necessary interventions such as dialysis. Refusing to treat the patient at all may be construed as patient abandonment. Although undiagnosed mental illness could be one factor in this patient's decision making, the physician must first determine the reasoning behind the patient's choices before referring the patient to a psychiatrist. Please refer to Case 10 for further discussion about when patients make decisions that their provider considers illogical.

20. **A.** Although ideally one would hope that families would come together and make a unanimous decision in such a situation, the reality is that individual disagreement is relatively common and unanimity may never be reached. Based on laws determining the hierarchy of surrogate decision makers in most states, the children have the responsibility of speaking for this patient and "vote counting" that includes others or seeking a legal guardian are not appropriate options. Including other family members in the discussion to assist the children with their choices is sometimes helpful, although this can also add more conflicting opinions to the mix. Currently 12 states do not offer a priority list of health care decision makers, and individual state laws should always be consulted. Please refer to Case 16 for further discussion about surrogate decision makers and when next-of-kin are not unanimous in their choices.

21. **A.** Prescription drug abuse is a serious problem, and establishing guidelines within a medical practice for the safe prescription of controlled substances is common and ethical if done in an appropriate manner. A patient who is not agreeable to a reasonable contract should not be given the option for treatment without a contract if other patients in the practice have to abide by one. Refusing to treat the patient at all may be construed as patient abandonment, so referral to a specialist for a second opinion is likely the best choice here. Please refer to Case 9 for further discussion about narcotic prescriptions and the ethics of narcotic contracts.

22. **A.** Providing safe and efficient care to this patient is a challenge and, unfortunately, establishing a behavioral contract may be necessary. Discharging the patient is not a good choice since he still requires inpatient treatment, and outpatient management of complicated pancreatitis is not standard of care. Complete

confinement to his hospital room or constant observation by a staff member is too restrictive of his rights at this time, and threatening the patient with discharge is not appropriate. Please refer to Case 5 for further discussion about patients whose behavior limits the ability of their providers to care for them, and the ethics of behavioral contracts.

23. **C.** Withholding medical resources from a patient generally should not be done on an individual basis, particularly if those resources are plentiful. When resources are limited one must consider the needs of other patients when distributing those resources. If care for Patient A is futile, then consideration of limiting access to Medication X should be made. Age and other factors, such as contribution to society, are not appropriate standards to use for limiting a patient's access to lifesaving treatments. Please refer to Case 22 for further discussion about determining how best to distribute medical resources among patients, and Case 14 about how age is a factor in health care decisions.

24. **B.** Immediate and completely honest admission to errors that may have affected a patient's care is essential, and should be done by the treating physician, preferably in person. Please refer to Case 17 for further discussion about how to deal with errors involving patient care.

25. **D.** In certain emergency cases informed consent may be waived if the patient's life is hanging in the balance. Clear documentation of this should be made in the medical record. If there is reason to believe that the patient's husband may have caused harm to the patient, then he should not be considered an appropriate surrogate decision maker for her. There is not enough time in this case to find an alternate surrogate decision maker for the patient, such as her mother or a legal guardian. Please refer to Case 4 for further discussion about when a patient's legal spouse is not an appropriate surrogate decision maker for them, and to Case 15 for a review of informed consent in emergency situations.

26. **A.** The individual certifying the death certificate in this case needs to do so with the best of their ability and using complete honesty. If the primary cause of death is known to be AIDS, then this must be listed and a second opinion from the medical examiner is not necessary. The role of a patient's health care agent is to make treatment decisions for them based on what the agent perceives as being in the patient's best interest or what the patient would want, and completing a death certificate is not a treatment decision. Please refer to Case 21 for further discussion about how to complete death certificates, and how to list illnesses with potential social stigma.

27. **B.** Exposing children to the manufacture or distribution of illicit drugs is largely construed as child abuse and this must be reported to the department of social services to ensure that the children are in a safe home situation at all times.

Urine toxicology screens that are analyzed for the purpose of medical care should not be used to prosecute patients. Notifying the patient's ex-wife of the test results would be a breach in patient confidentiality. Please refer to Case 7 for further discussion about what determines child abuse and the obligations of providers to report suspected abuse.

28. **C.** Informed consent is preferable, but not always necessary in the case of life-threatening emergencies, and this patient requires emergent central venous access for her care. Documentation in the medical record is essential when a procedure is performed on a patient without their consent. Waiting for a legal judgment is not appropriate in an emergency situation. Please refer to Case 15 for further discussion about informed consent in emergency situations, and consent in unidentifiable patients.

29. **D.** Most states have regulations that require formal informed consent before ordering certain specific genetic assays. Although in most states the law dictates that HIV testing requires informed consent, no laws exist regarding formal consent for testing of surrogate laboratory markers for HIV infection. Nonetheless, positive results from these surrogate markers may still be reportable and the most ethical thing to do when ordering them is to notify the patient that they are being drawn and preferably obtain their consent to do so. Technically, however, consent for ordering these tests falls under general consent for medical care, as does testing for viral hepatitis. Please refer to Case 19 for further discussion about general consent for medical care, informed consent for HIV testing, and the use of surrogate laboratory markers for HIV infection.

30. **C.** This case represents a common clinical dilemma. Withholding medical resources from a patient generally should not be done on an individual basis, particularly if those resources are plentiful. Factors, such as underlying illness, may or may not be an appropriate standard to use for limiting a patient's access to lifesaving treatments, and these situations must be assessed on a case-by-case basis. If medical evidence suggests that an intervention may cause a patient more harm than good, it is the obligation of the practitioner to provide this information to the patient and their family. Additionally, if the underlying illness makes the intervention futile, then it probably should not be offered to the patient at all. Referral to a physician with the same race is both unnecessary and inappropriate. Please refer to Case 13 for further discussion about determining the futility of care, and how to approach end-of-life issues when cultural differences exist.

31. **A.** When a patient does not have the capacity to make a health care decision on their own, choices about the use of life-sustaining interventions are based on the patient's wishes which, preferably, have been made clear verbally or through documentation such as a Living Will. If evidence of the patient's wishes is not available, and there is no Living Will, then the health care agent from the Power

of Attorney for Health Care is the highest on the hierarchy for surrogate decision making. This is the person who the patient felt, at a time when they had capacity, could best make decisions for them in such a situation. There may be a reason that the patient in this case has chosen someone other than their legal spouse to be their health care agent (e.g., separation from the spouse, a dysfunctional or abusive relationship with the spouse, lack of capacity by the spouse) and so the health care agent has priority in this case. However, further information about the reasons for this patient's choice should probably be gathered and consideration made to obtaining legal consultation if the issue becomes a contentious one. Please refer to Case 2 and Chapter 3 for further discussions about determining who makes health care decisions for patients without the capacity to do so.

32. **D.** In situations where patients have not designated a health care agent for themselves, a surrogate decision maker must be chosen for them. This individual must have the capacity to make appropriate decisions that are in the best interest of the patient, and the health care team has to assess this capacity to their best ability. Barring any reasonable evidence that precludes the father from assuming this role in the presented case, the next best answer would be "C." If the father is already aware of the HIV diagnosis, then free discussion of this is permitted. However, if he is not aware of his son's HIV infection, this specific information should not be disclosed as it is not directly relevant to the care of his son's stroke. Nonetheless, all information that is relevant to care of the stroke must be completely disclosed so informed decisions can be made for the patient. Please refer to Case 11 for further discussion about reporting HIV diagnoses to family members of patients, and both Cases 2 and 11 about surrogate decision makers with potential conflicts of interest.

33. **D.** It is appropriate to discuss this procedure with a 100-year-old patient, providing honest evidence about the potential risks and benefits, to best determine if she wishes to proceed with having it done. This is assuming that the patient has the capacity to make such a decision, and if she does not then a surrogate decision maker will need to become involved. If performing a colonoscopy is considered completely futile in this case, then it should not be offered to her at all. Although it is true that this patient is beyond the current life expectancy of a person in the United States, decisions about potentially lifesaving procedures should be made on an individual basis and not based on a physician's preconceived notions of what an older patient might want or what their pre-hospital quality of life might be. This is true even if providing the procedure to the patient may not significantly prolong the duration of her life. Withholding medical resources from a patient generally should not be done on an individual basis, particularly if those resources are plentiful. The ability of a patient to pay for a potentially lifesaving procedure should not limit her access to it. Please refer to Case 14 for further discussion about age-related decisions in medicine.

34. **C.** A recent trend is to allow family members to be present during resuscitation efforts of their loved ones, as long as they do not interfere with the resuscitation efforts of the code team. This would require limiting the family members present to a reasonable number, and to ensure that these chosen members are adults that are emotionally capable of witnessing the event. Having the discussion about family member presence should be done in advance of the resuscitative efforts whenever possible, as it is in this case. Although some warning as to what the family may observe during resuscitation is prudent, descriptive words should be chosen wisely. Please refer to Case 18 for further discussion about family presence during resuscitation efforts.

35. **D.** Patients have the right to refuse medical care, even when it seems medically necessary to save their life. The risks and benefits of accepting or refusing the procedure must be explained to and understood by the patient in order for them to make an informed decision about such a refusal. If they show good understanding of the risks and benefits, then they have the capacity to decide. Bullying them into changing their mind is inappropriate, although they should be made aware that the procedure will likely still be made available to them at a later date should they voluntarily change their mind about having it done. A psychiatry consult may be helpful in this case if there is concern that an underlying mental illness is affecting the patient's judgment, but is not necessary in determining the patient's capacity to make this decision. Please refer to Case 1 for further discussion about determining the capacity of a patient to give informed consent.

36. **C.** Providers must make a good effort to accommodate the belief systems of patients with different cultural views, and not try to interfere with or change these beliefs. Holding insulin during periods of fast seems like a reasonable choice for this patient. Withholding insulin treatment for the majority of the year does not make practical sense for her illness. Many religious groups do in fact exempt patients from fasting if they are pregnant or have acute or chronic illness, and using a culturally-competent tone and recommending that the patient discuss the matter further with a religious advisor would be fine. Outright advising the patient to go against her belief system is not appropriate. Good care should be able to be provided by the physician, regardless of their own religious or cultural background. Please refer to Case 23 for further discussion about caring for patients with different belief systems, as well as Chapter 1, Case 16, and Case 23 on how religion and culture can affect medical treatment.

37. **D.** Withholding medical resources from a patient generally should not be done on an individual basis, particularly if those resources are plentiful. When resources are limited one must consider the needs of other patients when distributing those resources. If ECMO is available at a nearby facility, and transporting the second infant to that facility will not do them harm, then this seems like the best choice for the situation. Although cost is always a consideration in the care of all patients, it is more a matter of using the least expensive, yet equally

effective, treatments available. Factors such as contribution to society are not an appropriate standard to use for limiting a patient's access to lifesaving treatments. If withdrawal of ECMO is considered in this case, it should be because use of this intervention has become futile or is doing the patient more harm than good. Please refer to Case 22 for further discussion about caring for patients whose care impacts the availability of limited medical resources.

38. **C.** Although patient information must be kept confidential in most situations, confidentiality can be broken in circumstances where harm of others can be prevented. Purposeful transmission of HIV infection to others is considered criminal in many states, but it is not the first priority of the caring physician to have the patient immediately arrested. Trusting that this patient will immediately notify his wife of his HIV infection and stop having unprotected intercourse with her seems naïve based on his recent behavior. Positive HIV test results, in most states, are by routine reported to the local health department, and a counselor/representative from that department will interview the infected patient and notify their sexual partners of their risk for becoming HIV-infected. Please refer to Case 12 for further discussion about reportable illness and notification of persons at risk for contracting contagious illness from a patient.

39. **B.** The use of chemical or physical restraints is sometimes necessary in the care of patients to reduce harm to the patient or to the providers caring for them. The minimal amount of restraint should be used, and the need for continued restraint re-evaluated frequently. Waiting for the psychiatry evaluation, although one is clearly indicated for this patient, does not seem practical at this time due to the acute level of patient agitation and his behaviors. Verbal redirection will unlikely be successful in calming this patient, although placing him in a more quiet environment may help with his agitation and cause less of a disturbance to other patients. If the patient receives sedative medications and is placed in a secluded area, close observation will be necessary in case oversedation or an adverse reaction to the medications occurs. Please refer to Case 6 for further discussion about treating agitated patients and the use of physical and chemical restraints.

40. **B.** Although the majority of states (47) use confidential yet name-based reporting, three states (Hawaii, Maryland and Vermont) use a code-based reporting system of positive HIV test results to the CDC. It is important to note that, in 42 states, there are also regulations mandating the reporting of $CD4^+$-lymphocyte count results, and 43 states require the reporting of HIV viral load test results. Please refer to Case 19 for further discussion about reporting positive HIV tests and surrogate markers for HIV infection to the CDC.

41. **A.** Immediate and completely honest admission to errors that may have affected a patient's care is essential, and should be done by the treating physician.

Although doing this in person is preferable, a phone call is the fastest method of the choices listed and offers more direct communication and opportunities to address questions and concerns from the patient than a letter. Please refer to Case 17 for further discussion about disclosing medical errors.

42. **C.** A provider should not offer to prescribe alternative treatments that are not standard of care just because a patient requests them. In this case there is not yet a clear reason to withhold all treatments from the patient, since the patient is agreeable to taking certain medications. In fact, refusing to treat the patient at all may be construed as patient abandonment. Although undiagnosed mental illness could be one factor in this patient's decision making, the physician must first try to figure out the reasoning behind the patient's choices before referring her to a psychiatrist. Please refer to Case 10 for discussion about patients who make medically questionable decisions, and also Case 1 for a review on determining capacity.

43. **C.** Providing palliative sedation is physically different from giving a patient a prescription for medication that she can use to end her life, as is done in physician-assisted suicide. The latter is illegal in most states, and pushing morphine for the sole specific purpose of hastening death is illegal in all. The purpose of palliative sedation is to give the patient just enough medication to control her pain even though it may also hasten her death by depressing respiration. Relief of pain and suffering in a patient who is dying from a terminal illness may allow her to have a more peaceful and humane death. Please refer to Case 25 for further discussion about pain management at the end of life and the use of palliative sedation.

44. **C.** The hospital is obligated to make a good faith effort to arrange a safe discharge plan for this patient with a new physical disability. Leaving a patient with a new lower limb amputation alone in a second-floor apartment is not a good choice. The providers should not break confidentiality and share her medical information with relatives, although they might be a good resource for assistance here if she grants the medical team permission to speak with them. The patient no longer meets criteria for hospitalization and a timely discharge is appropriate. Working with her to decide upon a discharge plan that is amenable to both parties is the ideal solution. A truly difficult patient who refuses all assistance, even with a physical disability, may have to be discharged even without an appropriate plan in place. However, such decisions should probably be done with the aid of legal counsel. Please refer to Case 8 for further discussion about difficult discharges, and discharge planning for patients with disabilities.

45. **A.** Based on laws determining the hierarchy of surrogate decision makers in the majority of states, and since there is no legal spouse (who would take higher precedent), the patient's adult child has the responsibility of speaking for this patient. The hierarchy then follows, in order, the parent, the adult sibling and

others who are well-acquainted with the patient and their wishes. Currently 12 states do not offer a priority list such as this, and individual laws should always be consulted. Please refer to Case 16 for further discussion about determining the hierarchy of possible surrogate decision makers for the majority of states.

46. **D.** Discussions about end-of-life wishes are often difficult, particularly in patients with newly diagnosed terminal illness, and sensitivity must be used while initiating them. However, patients with potentially terminal conditions should have their wishes regarding end-of-life care clearly documented in the medical record and supplemented with legal documentation whenever possible. This patient should determine who he would want to make health care decisions for him if he loses the capacity to do so. His wishes regarding resuscitation, mechanical ventilation, and nutrition are some of the other parameters that should be clarified. Please refer to Cases 4 and 16 for further discussion about assignment of surrogate decision makers, and Chapter 3 for a review of Advance Directives.

47. **B.** In this patient, a moderately high risk surgery was performed two years previously, but ultimately symptoms returned due to patient non-adherence with medical recommendations. A second opinion will probably not change the diagnosis or treatment options here. Repeat surgery every two or so years is not an appropriate use of medical resources when the need for surgery is easily preventable with medical management, and may additionally place the patient at an unacceptably high risk. Outright refusal to operate again is the other extreme. A middle-of-the-road solution, perhaps more acceptable to all parties involved, would be to offer surgery after the patient has reasonably demonstrated adherence to medical recommendations. Such an effort will be important to the continued success of the surgery, risk reduction for the procedure itself, and to improve her general health overall. Non-adherence to medical therapy can occur for many reasons (e.g., cost, history of depression, etc.) and these reasons should be thoroughly investigated by the provider. However, there is no clear indication for a psychiatry evaluation at this time. Please refer to Case 20 for further discussion about determining plans of care in patients who are non-adherent to medical recommendations.

48. **A.** Patients who are legal minors can often make their own decisions about health care without involving their parents. The Code of Ethics of the American Medical Association suggests that providers delve into why minors do not want their parents involved in their care, and to encourage parental involvement when appropriate. This patient has stated that she is not yet ready to involve her parents in her situation, but the window is open to do this at a future date and a provider can offer to be present for such a discussion to help mediate or offer recommendations. Placing the patient on the spot to discuss the dilemma now is not appropriate, either alone or with assistance. Maintaining confidentiality,

without making false statements ("everything is fine") requires finesse, but is essential. Please refer to Case 24 for further discussion about caring for patients that are legal minors.

49. **D.** The provider here cannot divulge the cancer diagnosis to the daughter without the permission of the patient. Despite the stress induced by learning a cancer diagnosis, it is mandatory that the physician provide the biopsy results to the patient in a timely manner so that treatment options can be reviewed and a treatment plan initiated. Deferring to others for release of the biopsy results is not appropriate. Please refer to Case 21 for further discussion about patient confidentiality and withholding information from patients, and Case 4 for a review on the federal Health Insurance Portability and Accountability Act (HIPAA) Privacy Rule.

50. **C.** Although patient information must be kept confidential in most situations, confidentiality can be broken in circumstances where immediate harm of others can be prevented. If the patient is truly not sexually active with her husband, his risk of contracting gonorrhea is essentially zero. The patient should be the one who discusses the situation with her husband, but it is a priority to inform her about the required notification of contagious disease risk by the local public health department. Trusting that this patient will immediately notify her husband of the gonorrheal infection is not necessarily reliable. Although it appears that this patient's marriage is challenged, a recommendation of divorce does not take priority here. Please refer to Case 12 for further discussion about reportable illness and notification of persons at risk for contracting contagious illness from a patient.

Glossary of Terms in Medical Ethics

Advance Directives: State-authorized documents in which a capable person makes his or her wishes for health care known at a time before he or she is no longer able to make or communicate these decisions. Advance directives include the Living Will and the Power of Attorney for Health Care.

Autonomy: The capacity for self-determination; the capacity to make one's own decisions, based on the processing of truthful information, without pressure or undue influence.

Best Interests Standard: A standard of surrogate decision making used when health care agents and other surrogates do not know what the incapable patient would have wanted. Using this standard, they must make health care decisions based on what would be in the best interests of that patient. This standard considers the benefits and risks of medical interventions and the patient's quality of life.

Capable: A patient is capable when he or she is able to process information and make medical decisions based on that information.

Confidentiality: The requirement that medical information about a patient must not be revealed to anyone who is not involved in the care of that patient.

Futile Care: See 'Ineffective Interventions.'

Health Care Agent: A person who is appointed by the patient to make health care decisions for the patient when the patient is no longer able to make or communicate these decisions; may also be termed "representative," "surrogate" or "proxy."

Ineffective Interventions: Medical procedures that will not be effective in meeting any reasonable goals for the patient; also termed "futile care."

Living Will: A legal document expressing the desire that medical technology not be used to prolong the dying process. The Living Will goes into effect when the patient is no longer able to make or communicate health care decisions.

Power of Attorney for Health Care: A legal document in which a capable patient appoints a health care agent to make medical decisions when the patient is no longer able to make or communicate such decisions. In some states this document is referred to as the "Health Care Power of Attorney" or the "Durable Power of Attorney for Health Care."

Principle of Beneficence: Medical practitioners should act in the best interests of the patient. More specifically, they should prevent harm, remove harm, and promote good for the patient.

Principle of Distributive Justice: Health care resources should be distributed in a fair way among the members of society.

Principle of Non-Maleficence: Medical practitioners must not harm the patient.

Principle of Respect for Autonomy: Capable patients must be allowed to accept or refuse recommended medical interventions.

Principle of Respect for Dignity: Patients, their families, and surrogate decision makers, as well as health care providers, all have the right to dignity.

Principle of Veracity: The capable patient must be provided with the complete truth about his or her medical condition.

Proxy: See 'Health Care Agent.'

Representative: See 'Health Care Agent.'

Substituted Judgment: A standard of surrogate decision making used when the surrogate is able to make a reasonable judgment about what the patient would want. This could be based on conversations with the patient, the patient's verbally expressed desires, or the patient's relevant values and beliefs.

Surrogate: See 'Health Care Agent.'

Surrogate Decision Maker: A person appointed in a Power of Attorney for Health Care or recognized by state law or common law as authorized to make medical decisions for the patient when the patient is no longer able to make or communicate such decisions.

Voluntary Informed Consent: The requirement that, prior to undergoing recommended medical procedures, capable patients must agree to these procedures based on full, relevant, and truthful information that the patient understands, without coercion or undue influence. The information that should be provided includes the expected benefits and potential risks of the recommended procedure and any viable alternatives, as well as the results of refusing treatment altogether.

Index

A

Abandonment, patient. *See* Patient abandonment
Abortion, 1, 4, 7, 67, 188–192
Abuse, of children. *See* Child abuse and neglect
ACLS. *See* Advanced cardiovascular life support
ACP Ethics Manual. *See* American College of Physicians Ethics Manual
Acute physiology and chronic health evaluation (APACHE) scoring, 96, 133
Acute respiratory distress syndrome (ARDS), 8, 94, 96
Adolescent patients. *See* Minors and children as patients
Advance directives, 17–20, 33, 62, 98, 115, 220, 222, 230, 233
Advanced cardiovascular life support (ACLS), 12, 32–33, 35, 88–90, 100, 112–113, 146, 170
African-American patients, 3, 6, 95, 113, 136
AIDS. *See* HIV infection and AIDS
Alzheimer's disease, 13–14, 18, 31, 34, 171
AMA Code of Ethics. *See* American Medical Association Code of Ethics
American Association of Critical Care Nurses, 146
American College of Physicians (ACP) Ethics Manual, 3, 115, 122, 146, 169, 177, 197–198
American Council of Pharmaceutical Education (ACPE), 6
American Heart Association, 33, 88, 146
American Medical Association (AMA) Code of Ethics, 2, 115–116, 169, 190, 197–198, 220, 230
American Medical Student Association (AMSA), 8
APACHE scoring. *See* Acute physiology and chronic health evaluation scoring

Apologizing for mistakes, 142, 208–209, 214
ARDS. *See* Acute respiratory distress syndrome
Artificial nutrition and hydration, 4, 12, 17, 112, 126, 196, 222, 230
Assisted suicide, 7, 196–198, 205, 221, 229
Association of American Medical Colleges (AAMC), 5
Asthma, 65–67, 72
Atrial fibrillation, 59, 61, 120, 131, 133
Autonomy, Principle of Respect for. *See* Principle of Respect for Autonomy

B

Beauchamp and Childress, 3, 11
Behavioral contracts, 54–57, 204, 208, 220, 223–224
Belmont Report, 3
Beneficence, Principle of. *See* Principle of Beneficence
"Best interests standard," 19, 127–128, 233
Blood products, cost associated with, 175
Blood product transfusion, 87–90, 174–178, 181–184, 205
Blood substitutes, 183–185
Bloodless surgery, 182, 184
Breast cancer, 47–48, 202, 215, 217

C

Cancer, breast. *See* Breast cancer
Cancer, ovarian. *See* Ovarian Cancer
Cancer, pancreatic. *See* Pancreatic cancer
Capacity for decision making, 2, 5–7, 13–14, 17–19, 25–30, 32, 34–35, 62, 88–90, 96–97, 101, 116, 129, 154, 168, 170–171, 219–222, 225–227, 229–230, 233

Cardiopulmonary arrest, 4, 25, 32–34, 88, 112,
    117, 120–121, 132, 145, 148, 210–213
Cardiopulmonary resuscitation (CPR). *See*
    Advanced cardiovascular life support
CD$_4$$^+$-lymphocyte counts, 87, 93, 96, 152,
    154–155, 203, 210, 214, 228
Centers for Disease Control and Prevention
    (CDC), 48, 95–96, 104–105, 113, 133,
    152, 154, 168, 188, 214, 228
Child abuse and neglect, 65–70, 72–73, 111,
    135, 189, 224–225
Child Abuse Prevention and Treatment
    Act (CAPTA), 67
Children as patients. *See* Minors and children
    as patients
Cholecystitis, 181–184
Chronic pain, 77, 83
Cocaine, 53, 65–67, 70, 73, 75, 82, 151, 167, 210
Coma, 14, 114, 126, 131, 133
Commission on Collegiate Nursing Education
    (CCNE), 6
Competence. *See* Capacity
Competency. *See* Capacity
Confidentiality, 1, 14, 49, 69, 97–100, 104,
    107–109, 154, 157–158, 187–191, 210,
    213–214, 216, 225, 228–231, 233
Contracts, behavioral. *See* Behavioral contracts
Contracts, narcotic. *See* Narcotic contracts
Core clinical competencies, 23
Cost of care, 4, 8, 22, 27, 41, 43, 77, 115,
    120–122, 142–143, 162, 168, 178, 208,
    213, 227
Court cases. *See also* United States Supreme
    Court and Court of Appeals decisions
    Cobbs v. Grant, 169
    *In re the conservatorship of Helga M
        Wanglie*, 113–114
    Public Health Trust v. Wons, 183
    *Tarasoff v. Regents of the University of
        California*, 69, 107
CPR. *See* Advanced cardiac life support
Cultural competence, 2, 5–7, 11, 14, 112, 116,
    135–136, 219, 225, 227
Culture, 1, 5–7, 23, 75, 136, 219, 227

D
Death certificates, 168–171, 209, 224
Death with Dignity Act (Oregon), 197–198
Defibrillation, *See* Advanced Cardiovascular
    life support
Delirium, 60, 62
Dementia, 14, 18, 25, 30–32, 36, 60, 203–204
Dignity, Principle of Respect for. *See* Principle
    of Respect for Dignity

Disabilities, 12, 41, 77–79, 229
Discharge of a patient
    administrative, 54–55, 57, 81–85, 206,
        208, 222
    from the hospital, 18, 33, 39–44, 55, 57,
        75–79, 81–85, 161, 201–202, 204, 208,
        215, 219–220, 222, 224, 229
Disclosure of diagnoses to patients, 94, 107,
    169, 208–209, 211, 214, 217, 226
Disclosure of errors or mistakes to patients,
    140–142, 208, 209, 214, 224, 228–229
Disruptive patients, 54, 147
Distribution of illicit drugs, 65, 68–70, 224
Distributive Justice, Principle of. *See* Principle
    of Distributive Justice
Divorce, 48, 167, 231
DNR orders. *See* Do Not Resuscitate orders
Domestic partnerships. *See* Same-sex
    partnerships
Do Not Resuscitate orders, 12, 33, 170, 174,
    196, 205
Drug abuse, 70–71, 87, 167, 173
    treatment of, 56, 71, 223
Drug dealing. *See* Distribution of illicit
    drugs
Drug screens and testing. *See* Urine toxicology
    screens
Drug-seeking behavior, 82, 85
Drug use in pregnancy, 67
Durable Health Care Power of Attorney. *See*
    Power of Attorney for Health Care

E
Emergency contraception, 4, 7, 188
Emergency Nurses Association, 146
Endocarditis, 53–55, 204
End-of-life care and hospice, 15, 17, 35, 41,
    48–50, 94, 99, 132, 134–136, 146, 169,
    170, 195, 197, 219, 221–222, 225,
    230
End-Stage Liver Disease (ESLD), 173–174,
    176–177
End-Stage Renal Disease (ESRD), 41,
    161–162, 164
Ethics
    definition, 1
    education, 5–6, 21
    in medical education, 1–9, 11, 21–23
    in nursing education, 6, 21
    in pharmacy education, 21
Ethnicity, 23, 132, 136
Euthanasia, 1, 197–198
Evidence-based medicine as a topic, 8–9, 22,
    221

F

Family
definition of, 132
involvement, 5–6, 8–9, 12, 14–15, 18–20,
33–34, 40, 42–44, 49–50, 56, 62, 84,
90–91, 94, 96, 100–101, 112, 141, 164,
191–192, 220–221, 230
presence at codes, 146
Feeding tubes. *See* Percutaneous endoscopic
gastrostomy (PEG) tubes
Femoral fractures, 25–26, 28, 125
Fourteenth Amendment of the U.S.
Constitution, 189
Fourth Amendment of the U.S. Constitution,
70
Futility, concept of medical, 6, 8, 36, 48, 50,
94, 96, 112–118, 121, 128, 133, 135,
147, 177–178, 202, 219, 223–226,
228, 233

G

Gastrointestinal bleeding, 87–90, 94–95, 105,
173–177, 212
Gay patients. *See* Homosexual patients
Glasgow Coma Scale (GCS), 126, 131

H

Health and Human Services. *See* United States
Department of Health and Human
Services
Health care agent, 17–18, 20, 25–27, 31–36,
48–49, 97, 113, 126, 132–134, 203,
209, 211, 220, 222, 224–226, 233–234
Health Care Financing Administration, 55, 61
Health Care Power of Attorney. *See* Power of
Attorney for Health Care
Health Insurance Portability and
Accountability Act of 1996 (HIPAA),
49, 97, 107, 222, 231
Hepatitis testing, 72, 151–153, 165, 210, 225
Heterosexual transmission of HIV. *See* HIV,
transmission routes of
HHS. *See* United States Department of Health
and Human Services
Hispanic patients, 113, 136
HIV
and organ transplantation candidates, 162,
213
consent for testing, 71, 73, 103, 151–154,
157–158, 167, 188, 190, 201, 203,
210, 219

disclosure of diagnosis, 94–97, 168, 171,
209, 211, 213, 226, 228
infection and AIDS, 87, 93–96, 99–100,
104–107, 152–154, 157, 159, 162,
168–169, 224
post-exposure prophylaxis (PEP)
for, 152
reporting of diagnosis, 104–108,
154–156, 168–171, 209, 214,
220, 225, 228
risk factors for, 72, 95, 105, 107, 169, 220
surrogate laboratory tests for, 152, 154,
157–158, 210, 214, 220, 225, 228–229
testing in pregnancy, 7, 70
transmission routes of, 95, 105, 152, 168
treatment for, 95–96, 104–105,
107–108, 152, 157–158, 165
Homosexual patients, 94–95, 99, 105
Hospice care, *See* End-of-life care and hospice
Hospital policies and regulations, 16, 34,
41–44, 55, 57, 61–62, 83

I

Informed consent. *See* Voluntary informed
consent
Injection drug use (IDU), 55, 95, 168–169,
173, 201, 204
Institutional Review Boards (IRBs), 3
Intracerebral hemorrhage, 111, 114, 131, 133
Intubation, endotracheal, 25, 88–90, 94, 112,
125–126, 174, 196, 205
Involuntary commitment, 54, 57
Involvement of family. *See* Family
involvement
Irby, David, 21

J

Jehovah's Witnesses, 181–185, 205, 221
Justice, Principle of Distributive. *See* Principle
of Distributive Justice

L

Liaison Committee on Medical Education
(LCME), 5
Life support, decisions about, 8, 32, 35, 112,
132, 135–136
Litigation, 4, 40, 43–44, 143
Living Will, 17–19, 25, 27, 32, 36, 48, 51, 62,
112, 132, 196, 206–207, 211, 215–216,
225, 233

M

Malpractice, medical, 4, 114, 141–142, 206, 222

Mechanical ventilation, 3, 6, 8, 32–35, 88–90, 94, 96, 112–113, 117, 125–128, 132–133, 135, 202, 211, 222, 230

Medicaid, 55, 61, 120, 177, 183

Medical education. *See* Ethics, in medical education

Medical ethics, definition, 1

Medical necessity, 41

Medical University of South Carolina, 70

Medicare, 33, 41, 43, 55, 61, 77, 120–122, 162, 177, 183, 212

Men who have sex with men (MSM), 95, 105

Minors and children as patients, 4, 14, 30, 79, 95, 105, 168, 171–172, 187–192, 220, 230–231

Model for End-Stage Liver Disease (MELD) score, 173–175

Morning-after pill. *See* Emergency contraception

MSM. *See* Men who have sex with men

N

Narcotic contracts, 82–84, 207, 223

National Center for Health Statistics, 169

National League of Nursing Accrediting Commission (NLNAC), 6

Nationwide Blood Collection and Utilization Survey, 175

Natural death, concept of, 12, 17, 99, 135

Natural Death Act of California, 17

Neglect. *See* Child abuse and neglect

No Free Lunch program, 8

Non-adherence with medical recommendations or treatment, 55–57, 96, 161–165, 208, 213, 216, 230

Non-Maleficence, Principle of. *See* Principle of Non-Maleficence

Nuremberg Code, 2

Nursing education. *See* Ethics, in nursing education

O

Oath of Hippocrates, 1–2

Oath of Maimonides, 2

Oaths, 1–3

Oregon's Death with Dignity Act. *See* Death with Dignity Act

Ovarian cancer, 140–143

Oxygenation during cardiac arrest, 88–89

P

Pacemakers, 26–28, 119–122

Pain management, 83, 197, 207, 229

Palliative sedation, 197–199, 221, 229

Pancreatic cancer, 167–171, 195–196, 205

Pancytopenia, 87–88, 167

Paraplegia. *See* Spinal cord injuries (SCIs)

Partners, domestic. *See* Same-sex partnerships

Paternalism, 2, 57, 165

Patient abandonment, 55, 84, 206, 222–223, 229

Patient Self-Determination Act, 98, 183–184

Percutaneous endoscopic gastrostomy (PEG) tubes, 17, 112, 115–117, 126–128

Pharmaceutical company influence on health care providers and trainees, 7–8

Pharm-Free program, 8

Pneumocystis jirovecii pneumonia (PCP), 93–94, 96, 104

Power of Attorney for Health Care, 18–20, 25, 27, 31, 34, 51, 62, 94, 96–97, 112, 132, 134, 203, 206–207, 211, 215–216, 220, 233–234

Prenatal exposure to drugs, 68, 70

Pregnancy, 4, 7, 67, 70, 184, 187–192, 216, 227

  in Jehovah's Witnesses, 184

  in teenagers, 187–189, 192, 216

Principle of Beneficence, 3, 11–12, 19, 22, 28, 34, 42, 49–50, 56–57, 61, 70–71, 78, 90, 99, 108, 116, 121, 127, 135, 142, 147, 157, 163, 170, 176, 183, 190, 198, 234

Principle of Distributive Justice, 3, 11, 15, 16, 22, 43, 56, 78, 117, 122, 128, 164, 177, 234

Principle of Non-Maleficence, 3, 11, 12, 19, 22, 26, 28, 34, 42, 49, 61, 70–71, 78, 84, 90, 99, 108, 116, 121, 127, 135, 147, 157, 163, 170, 176, 183, 190, 198, 234

Principle of Respect for Autonomy, 3, 11, 13, 15, 17–19, 22, 28–29, 34, 42, 50, 56, 62, 71, 78, 84, 90, 99, 115, 121, 127, 135, 147, 157, 164, 169, 176, 184, 191, 199, 234

Principle of Respect for Dignity, 11, 13–15, 17–18, 22, 29, 35, 50, 56, 90, 99, 108, 116, 128, 135, 142, 147, 158, 164, 170–171, 177, 184, 191, 199, 234

Principle of Veracity, 11, 15, 22, 28, 42, 62, 99, 142, 171, 234
Principles of Biomedical Ethics, 3, 11
Protected health information (PHI), 49, 97
Psychiatry consults, 54, 90, 204, 214, 221, 227–228, 230
Public Health Service Act, 55, 83, 176–177

R
Race, 23, 132, 136, 225
Reimbursement for medical services, 41–42, 77
Religion and religious convictions, 4, 6–7, 11, 14, 23, 29, 101, 112, 116, 135–136, 183–185, 192, 205, 212–213, 219, 227
Reportable illnesses, 104–109, 154–159, 188–189, 191, 203, 210, 214, 220, 224–226, 228, 231
Reporting abuse or neglect of children, 65–72, 210
Resource allocation, 43, 122, 174, 177
Respect for Autonomy, Principle of. *See* Principle of Respect for Autonomy
Respect for Dignity, Principle of. *See* Principle of Respect for Dignity
Restraints, use of, 55, 57, 60–63, 204, 214, 221, 228
Resuscitative efforts, *See* Advanced cardiovascular life support

S
Same-sex partnerships, 94, 97–98, 101, 137, 222
Same-sex relationships, legal recognition of, 94, 97–98
Schiavo, Terri, 4–5
Seclusion of a patient, 55–57
Sexually-transmitted infections (STIs), 104–105, 187–190, 192, 220
Sexual orientation, divulgence of, 95, 101
Spinal cord injuries (SCIs), 75–77
Stroke, 18, 39, 111–114, 133, 135, 211, 226
indicators of futility, 113–114
"Substituted judgment," 19, 33–34, 115, 127, 134, 199, 234
"Sundowning," 60
Surrogate decision makers, 13–16, 18–19, 32–36, 49–51, 62, 94, 96–101, 114–116, 127, 133–134, 172, 199, 206, 209, 211, 214, 216, 222–224, 226, 229–230, 233–234

determining hierarchy for, 19, 97, 133-134, 220, 223, 226, 229–230
Syphilis, 2–3, 103–105, 107–108

T
Terminal extubation, 94, 99, 113
Terminal illness, 6, 15, 17–18, 27, 32, 167–169, 195–200, 211, 221–222, 229–230
Tetraplegia. *See* Spinal cord injuries (SCIs)
Texas Advance Directives Act, 115
Threats, 40, 43, 69, 100, 107, 206
Toxicology screens, urine. *See* Drug screens
Tracheostomy, 112, 115–117, 126
Transfusion of blood products. *See* Blood product transfusion
Transplant (organ) allocation, 41, 161–165, 174, 176–177
Tuskegee Syphilis Study, 2–3, 6, 136

U
Uniform Anatomical Gift Act, 163
United Network for Organ Sharing (UNOS), 163, 174
United States Department of Health and Human Services (HHS), 78, 105
United States Supreme Court and Court of Appeals decisions. *See* also Court cases
Canterbury v. Spence, 126, 169
Cruzan v. Director, Missouri Department of Health, 17
Ferguson v. City of Charleston, 70
Roe v. Wade, 189
Vacco v. Quill, 198
University of Massachusetts Medical School, 2
University of Washington School of Medicine, 6
University of Toronto, 8
Urine toxicology screens, 53, 65–67, 69–73, 75, 82–84, 187, 201, 207, 210, 219, 225

V
Ventricular fibrillation, 33
Veracity, Principle of. *See* Principle of Veracity
Vertigo, 39, 41
Visitors in the hospital, preventing. *See* Seclusion of a patient

Voluntary informed consent, 3, 13, 28, 49,
    70–71, 87, 122, 125–126, 151,
    153–154, 156–157, 159, 169, 184,
    191, 201–203, 209–210, 219–221,
    224–225, 227, 234

W
Withholding information from patients,
    168–171, 231